Noise in Radio-Frequency Electronics and its Measurement

Series Editor
Robert Baptist

Noise in Radio-Frequency Electronics and its Measurement

François Fouquet

iSTE

WILEY

First published 2020 in Great Britain and the United States by ISTE Ltd and John Wiley & Sons, Inc.

ISTE Ltd
27-37 St George's Road
London SW19 4EU
UK

www.iste.co.uk

John Wiley & Sons, Inc.
111 River Street
Hoboken, NJ 07030
USA

www.wiley.com

Library of Congress Control Number: 2019953869

British Library Cataloguing-in-Publication Data
A CIP record for this book is available from the British Library
ISBN 978-1-78630-532-9

MIX
Paper from responsible sources
FSC www.fsc.org FSC® C013604

Contents

Preface

As a professor/researcher at ESIGELEC since 1982, I have been confronted with "noise in radiofrequency electronics" as part of my teaching activities, as also during my research work.

For my teaching activities, it was in the second half of the 1980s, that I created course materials, tutorials and practical work on this subject for students of ESIGELEC pursuing courses in the field of radiofrequencies and microwaves and for technicians and engineers working for companies in these fields.

As far as my research activities are concerned, my first "in-depth" contact with "electronic noise" dates back to the time of my DEA diploma and then my doctorate, when I had to quantify "by hand" the impact of the use of "active load polarization" on the noise factor of a "common gate MESFET".

In both cases, I was confronted with the same problem: unsuitable bibliographic sources. The reason being either:

– we read course documents, where only the basics were presented and often briefly, sometimes omitting notation details, making the information difficult to use and where long demonstrations were omitted too because they can be a bother but could allow an understanding of what was really going on; or

– we read articles in scientific journals, and there the prerequisites were high level and the long demonstrations, vital for the correct use of the information, had to be reconstituted by the reader from the assumptions provided and the result given. It could take days, weeks or even months, depending on the reader's level.

It is to fill a part of the gap that exists between these two worlds that I decided to write this document to summarize what I have essentially learned about noise during the last 40 years.

On this occasion, I would particularly like to thank Mr. J.L. Gautier and Mr. D. Pasquet who initiated me to study this subject during my DEA diploma studies and my doctorate, while they were professors at ENSEA.

I would also like to thank Mr. M. Rivette and Mr. J.B. Dioux, alumni of ESIGELEC, who inspired me to do this work while I was correcting their report on the noise figure of adapted attenuator.

François FOUQUET
November 2019

List of Symbols

$*$	Conjugate operator on a complex, $c = a + jb, c^* = a - jb$
a_i	Incident power wave on port i of a two-port, equal to $a_i = \frac{v_i+i_i}{2}$ with $v_i = \frac{V_i}{\sqrt{R_0}}$ and $i_i = I_i \cdot \sqrt{R_0}$
b_i	Reflected power wave on port i of a two-port, equal to $b_i = \frac{v_i-i_i}{2}$ with $v_i = \frac{V_i}{\sqrt{R_0}}$ and $i_i = I_i \cdot \sqrt{R_0}$
b_{ni}	Noise power wave outputted from port i of a two-port
B_{Cor}	Imaginary part of the correlation admittance Y_{Cor}, between e and i of normalized value b_{Cor}
B_{Sopt}	Imaginary part of the optimal admittance for the Y_{Sopt} noise, of normalized value b_{Sopt}
e	Equivalent noise voltage source at the input of a two-port, see Figure 1.11
ENR	Excess noise ratio of a noise source at two temperatures T_C (cold temperature) and T_H (hot temperature), equal to $ENR = \frac{T_H-T_0}{T_0}$ and expressed in dB: $ENR_{dB} = 10 \cdot log_{10}(ENR)$
F	Noise factor of a two-port, information on the degradation of the signal / noise ratio between the input and the output of the two-port, depends on the load presented at the input of the two-port, $F > 1$

F_{eMin} The minimum excess noise factor of a two-port, equal to $F_{eMin} = F_{Min} - 1$

F_{Min} The minimum noise factor of a two-port, obtained for the optimal load for noise presented at the input of the two-port

Γ_{Sopt} Optimum reflection coefficient for the noise to present at the input of the two-port to obtain $F = F_{Min}$, equal to $\Gamma_{Sopt} = \frac{1-y_{Sopt}}{1+y_{Sopt}}$

$G_{Avai}(Y_S)$ Gain in Available power gain of a two-port for a source admittance Y_S, equal to the value of the transducer power gain $G_T(Y_S, Y_L)$ for $Y_L = Y_{OUT}^*$

G_n Equivalent noise conductance at the input of a two-port, $\overline{|\iota_n|^2} = 4kT \cdot df \cdot G_n$, of normalized value $g_n = G_n \cdot R_0$

G_{Sopt} Real part of the optimal admittance for the Y_{Sopt} noise, of normalized value g_{Sopt}

$G_T(Y_S, Y_L)$ Transducic power gain of a two-port for a source admittance Y_S and a load admittance Y_L, equal to the ratio of the power collected in Y_L and the power available at the source

G_0 Normalization admittance equal to $1/R_0$or $20mS$

h Planck's constant 6.23×10^{-34}J s

i Equivalent noise current generator at the input of a two-port, see Figure 1.11

$Im(c)$ "Imaginary part" operator on the complex c

i_n Equivalent noise current generator at the input of a two-port not correlated with e

i_1 Noise current generator equivalent to the input of a two-port, see Figure 1.10

i_2 Noise current generator equivalent to the input of a two-port, see Figure 1.10

j Square root of - 1 which allows to describe a complex in the form $c = Re(c) + j \cdot Im(c)$

k Boltzmann's constant 1.38×10^{-23} J K^{-1}

N_C	Noise power measured at the output of a two-port when the noise source is at cold temperature T_C
N_H	Noise power measured at the output of a two-port when the noise source is at hot temperature T_H
NF	Noise Figure of a two-port, equal to $10 \cdot log_{10}(F)$, $NF > 0dB$
μ_n	Electron mobility in $cm^2\ s^{-1}\ V^{-1}$
P	Power dissipated in a load Y, equal to $Re(V \cdot I^*)$ with $I = Y \cdot V$
P_{SAvai}	Available power of a source $E, Y_S = G_S + jB_S$, equal to $P_{SAvai} = \frac{\|E\|^2 \cdot \|Y_S\|^2}{4G_S}$
$Re(c)$	"Real part" operator on the complex c
R_N	Equivalent noise resistance at the input of a two-port, $\overline{\|e\|^2} = 4kT \cdot df \cdot R_N$, of normalized value $r_N = \frac{R_N}{R_0}$
R_0	Normalization resistance equal to 50 Ω
S_{ij}	S parameters of a two-port of ports i, j
T	Temperature in K
T_e	Equivalent noise temperature of a two-port equal to $T_e = T_0 \cdot (F - 1)$, expressed in K
T_{eMin}	Equivalent minimum noise temperature of a two-port equal to $T_{eMin} = T_0 \cdot (F_{Min} - 1)$, expressed in K
T_0	Reference temperature for the noise factor equal to 290 K
Y	"Y" factor, used by the measurement of the factor, equal to $Y = \frac{N_H}{N_C}$
Y_{Cor}	Correlation admittance between e and i, defined by $i = Y_{Cor} \cdot e + i_n$ or $Y_{Cor} = \frac{\overline{\|i.e^*\|}}{\overline{\|e\|^2}}$
Y_{ij}	Admittance parameters of a two-port of ports i, j
Y_{Sopt}	Optimum Admittance for the noise to present at the input of the two-port to obtain $F = F_{Min}$, of normalized value y_{Sopt}

Introduction

The purpose of this book is to provide precise information at the level of the notations and arguments that will allow the reader to:

– carry out literal calculations on an electronic circuit to predict noise behavior;

– interpret noise simulation results and rationally modify the circuit to improve its performance towards noise;

– conduct structured and consistent noise factor measurements or noise characterizations and have a critical eye on the results.

To arrive at this result, some demonstrations, sometimes long, will be detailed because it is a good way of understanding what a formula or an equation actually means when we are able to demonstrate the path which goes from the assumptions to the result; in particular, we really understand what each of the terms of this formula or equation is.

In terms of prerequisites, the level is that of an electronics engineer who knows Ohm's law, knows what a voltage or current generator is, whether free or controlled, and knows what impedance and power are in sine wave operation. Required are basic notions about complex numbers, but especially the desire to apply basic knowledge of electronics to build skills on the subject of noise in electronics that may seem, wrongly in my opinion, a difficult field.

The organization of the book is as follows:

– in Chapter 1, we will put in place the basics of noise in electronics such as the origins of background noise in electronics and the quantities used in this field with in particular the notion of noise factor and its importance in telecommunications;

– Chapter 2 is devoted exclusively to the Friis formula, which plays a key role in noise measurement, since no electronic device escapes background noise, not even the device that is used to measure the noise factor. This formula, being so fundamental, is sometimes so badly written that it loses all utility. We will re-demonstrate this formula with formalism that is much more legible than the one used in the original article by Friis (1944);

– in Chapter 3, we will focus on the case of passive devices that have a particular noise behavior since we can predict it from their small signal behavior. We will do it using an example – an adapted attenuator – before examining Bosma's theorem;

– in Chapters 4 and 5 we will see how to make measurements to determine, firstly, the noise factor on 50 Ω for a two-port adapted to 50 Ω and then how to completely characterize in noise any device. We will highlight the need for equipment, calibrations and calculations that are very different in both cases;

– some exercises are proposed in Chapter 6. They will allow the reader to apply the methods described in the previous chapters and to obtain new results, in particular on transistors.

In addition, the reader will find, in the appendices, useful information in the context of the noise on the admittance parameters, the S-matrix, the flow graph and Mason's rule and, finally, on the noise power waves. A reader unfamiliar with the admittance parameters and S-parameters will be interested in reading the first three appendices before reading this book.

It should be noted that there are no new results on noise in this study, apart from the few results from measurements made on the benches at ESIGELEC. On the other hand, the formulation of some problems has been revised to make them easier to understand for a reader unfamiliar with the field of background noise in electronics and to benefit from it.

Although literature is abundant on the subject of noise in scientific journals, the bibliography cited in this book is limited to what is strictly necessary to pay tribute to the pioneers in the field who, since the 1940s, have formalized a number of concepts, new at the time, but still relevant and useful. Nevertheless, some more recent articles are used, in particular on the problem of the noise in transistors in RFCMOS technology.

1

Background Noise in Electronics

1.1. Introduction

The objective of this chapter is to put in place the concepts necessary to quantify the impact of background noise on the processing of electrical signals by electronic functions. Indeed, noise is a key limiting factor in the field of electronic processing of low power information as is the case in telecommunications.

In an electronic system, noises can take different forms:

– noises of artificial origin, which are all related to human activity. These artificial noises can be qualified as spurious signals and are in fact electromagnetic disturbances conducted and/or radiated;

– noises of natural origin such as thermodynamic noise, atmospheric disturbances, solar activity and, more generally, radiation or cosmic radiation;

– noises related to spontaneous fluctuations in electronic components. These spontaneous fluctuations are permanently present in electronic components and are linked to mechanisms internal to these components.

In this study, we will focus only on the latter category because it is responsible for what is called background noise in the components. The first category is rather a problem of the type "electromagnetic compatibility" which does not fall within the thematic field of our analysis. Regarding the second category, even if there are obvious electronic manifestations of these noises (the generation of carriers per photon, the singular events), we will not take them into account in our study. In the rest of this chapter, we will describe the different mechanisms responsible for background noise, explain the concept of noise factor and its interest in telecommunications and finally model the noise in a two-port device (or "quadripole" if you prefer) and give expressions that allow to express the noise factor of this two-port and/or to characterize in noise a two-port.

1.2. Spontaneous fluctuations in electronic components

1.2.1. *Introduction*

The different mechanisms giving rise to background noise in electronics are:

– thermal noise;

– shot noise;

– generation / recombination noise;

– excess noise.

All these noises have in common a mean value of zero, but as they carry a certain power, they are characterized by their root mean square value which has the same definition for random variables as for deterministic signals.

In the rest of this section, we will describe these different noises by explaining the mechanisms involved.

1.2.2. *Thermal noise*

Thermal noise originates from the thermal agitation of free charges (conduction electrons and holes, where they exist) in metallic conductors or semiconductors. This noise is also called "Johnson noise", "Johnson-Nyquist noise" or "resistance noise".

The explanation of the phenomenon is the following: when a free charge – a conduction electron for example – jointly undergoes the deterministic action of an electric field $\vec{E}$ and the temperature T, its velocity is $\overrightarrow{v_{e-}} = -\mu_n \cdot \vec{E} + \overrightarrow{v_{Th}}$ where $\overrightarrow{v_{Th}}$ is the thermal velocity which is randomly distributed in module and argument. If one is interested in the current which results from the displacement of a set of electrons in a conductor and which is proportional to the projection of $\overrightarrow{v_{e-}}$ along the axis of the electric field $\vec{E}$, one finds instantaneous fluctuations that are related to the random distribution of $\overrightarrow{v_{Th}}$. This phenomenon is shown in Figure 1.1.

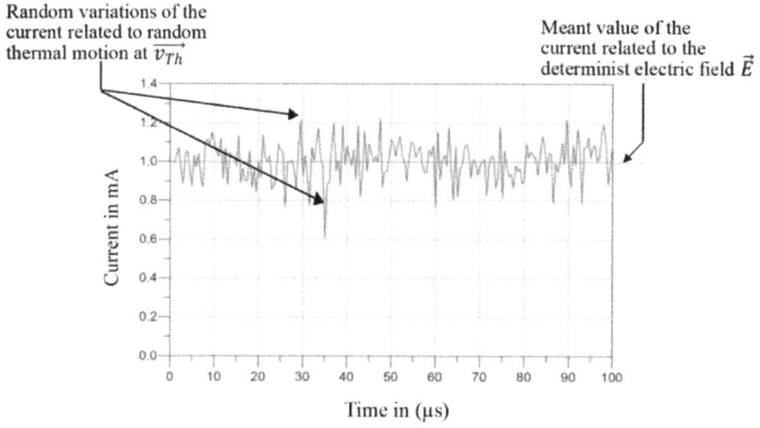

Figure 1.1. *Temporal representation of thermal noise*

Sources of thermal noise are resistors in all their forms:

– the “real” resistors;

– the non-active doped areas of semiconductor components, for example $R_{bb'}$, the base access resistance of a bipolar transistor as shown in Figure 1.2;

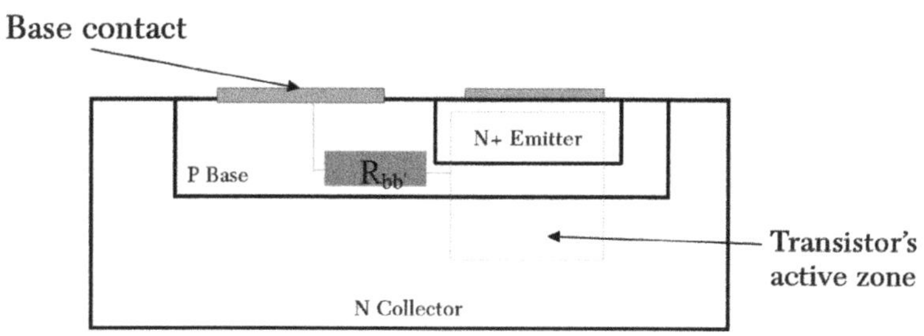

Figure 1.2. *Base access resistance of a bipolar transistor*

– the real part of the impedance of non-ideal reactive components such as the series resistance of an inductor or the loss conductance of a capacitor due to the real part of the dielectric constant of the insulation material used.

For low frequencies such that $f \ll \frac{kT}{h}$ where k is the Boltzmann constant, h is the Planck constant and T is the temperature of the resistor R, the power spectral density of the noise voltage V (that is $|\bar{V}|^2$ multiplied by R for 1 Hz bandwidth) is a constant equal to $2RkT$ that does not depend on the frequency considered: it is said that the thermal noise is a white noise.

As $h = 6.23 \times 10^{-34} Js$ and $k = 1.38 \times 10^{-23} J/K$, at $T = 300\, K$, it is sufficient that the frequency f is small compared to 6.6 THz. This situation is always verified at low frequencies and even at microwave frequencies. We can therefore use the Nyquist formula given by equation [1.1], which gives the root mean square value of the noise voltage v_n for a resistance R placed at temperature T:

$$v_n = \sqrt{4 \cdot kT \cdot df \cdot R}, \qquad [1.1]$$

where df is the frequency range considered.

From equation [1.1], the thermal noise can be modeled in a resistance R laced at the temperature T in Kelvin, as shown in Figure 1.3 by using:

– a resistance R at 0 K and a voltage generator v_n in series with $v_n = \sqrt{4 \cdot kT \cdot df \cdot R}$;

– a resistance R at 0 K and a current generator i_n in parallel with $i_n = \sqrt{4 \cdot kT \cdot df \cdot G}$ with $G = 1/R$.

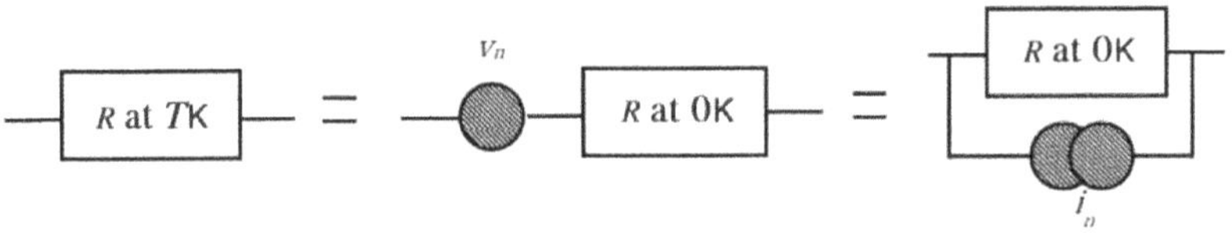

Figure 1.3. *Modeling a resistance for thermal noise*

Using the current model, one can ask the question of the available noise power across a conductance G placed at the temperature T K using the diagram of Figure 1.4.

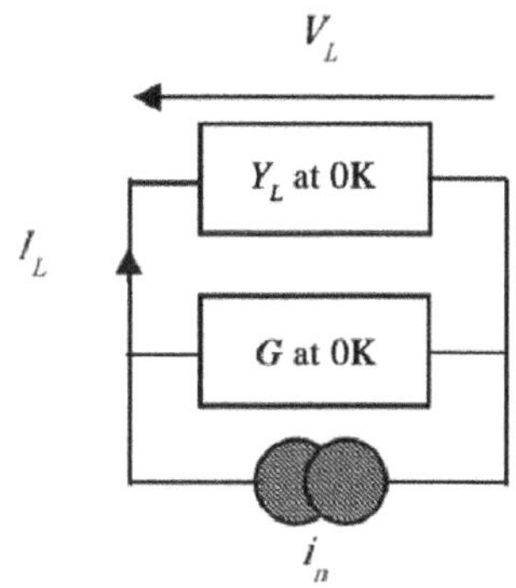

Figure 1.4. *Calculation of the available thermal noise power*

The value of this available noise power at terminals G is $kT.df$ is given by equation [1.2] and is obtained for $Y_L = G$:

$$P_{n,Avai} = P_L(Y_L = G) = |\overline{i_n}|^2 \cdot \frac{Re(Y_L)}{|Y_L+G|^2} = kT \cdot df \qquad [1.2]$$

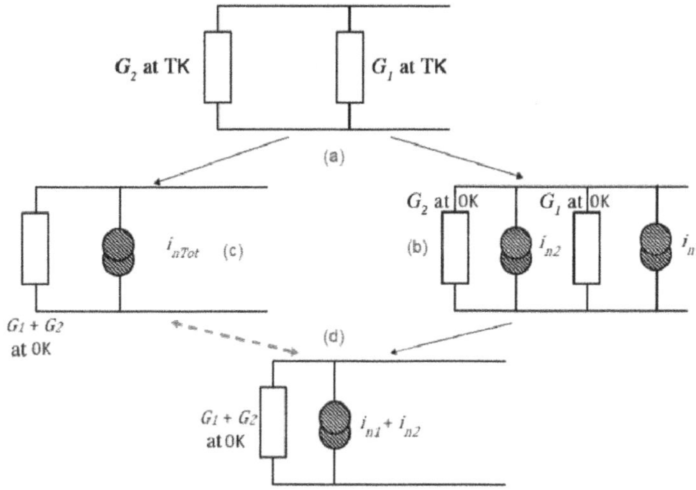

Figure 1.5. *Correlation between thermal noises*

We will now introduce an important concept when it comes to spontaneous fluctuations that behave like random variables: correlation. Consider Figure 1.5. The circuit of the top (a) comprises two conductances G_1 and G_2 placed at the same temperature T and connected in parallel. To model the noise, two possibilities can be envisaged:

– each conductance is modeled for noise and we obtain the circuit (b) of Figure 1.5 with $|\overline{\iota_{n1}}| = \sqrt{4 \cdot kT \cdot df \cdot G_1}$ and $|\overline{\iota_{n2}}| = \sqrt{4 \cdot kT \cdot df \cdot G_2}$;

– the equivalent conductance $G_1 + G_2$ is modeled for noise and we obtain the circuit (c) of Figure 1.5 with $|\overline{\iota_{nTot}}| = \sqrt{4 \cdot kT \cdot df \cdot (G_1 + G_2)}$.

From circuit (b), we can go to circuit (d) by adding the conductances and the noise generators. To be able to go from diagrams (c) to (d), the equality given by equation [1.3] must be verified:

$$|\overline{\iota_{nTot}}|^2 = |\overline{\iota_{n1} + \iota_{n2}}|^2 = |\overline{\iota_{n1}}|^2 + |\overline{\iota_{n2}}|^2 \qquad [1.3]$$

As random sources can be manipulated like complex quantities, we can write the equation [1.4]:

$$|\overline{\iota_{n1} + \iota_{n2}}|^2 = \left|\overline{(\iota_{n1} + \iota_{n2})(\iota_{n1}^* + \iota_{n2}^*)}\right| = |\overline{\iota_{n1}}|^2 + |\overline{\iota_{n2}}|^2 + 2.Re\left(\overline{\iota_{n1}.\iota_{n2}^*}\right) \quad [1.4]$$

For the equality [1.3] to be verified, the term $Re\left(\overline{\iota_{n1} \cdot \iota_{n2}^*}\right)$ of equation [1.4] must be zero. The nullity of this term means that the terms i_{n1} and i_{n2} are not correlated. To say that phenomena are not correlated means that there is no interaction between the two observed phenomena; which is conceivable quite well in our case because we do not see how the electrons of G_1 influence, by their movement of thermal agitation, the electrons of G_2 in their movement of thermal agitation.

When noise analysis of a circuit is done, it can always be said that two noise sources located in two distinct components are never correlated. On the other hand, within the same component, two sources of noise can be partially correlated. For example, a bipolar transistor, the base access resistance $R_{bb'}$ generates a base noise current which, by transistor effect, will generate a collector noise current independently of other phenomena giving rise to a collector noise current. Thus, in the end, if we represent the noise in a bipolar transistor by a base current generator and a collector current generator, then these two generators will be partially correlated.

1.2.3. *Shot noise*

Shot noise originates from the "discrete" nature of the current – each carrier contributes to the total current – and manifests itself in particular in semiconductor junctions. In a semiconductor junction, carriers must cross a potential barrier for an excitation that is randomly distributed around an average set, for example, by a polarization. If the carrier has received sufficient energy to cross the barrier, it passes and contributes to the current in the junction, otherwise it does not contribute.

This noise is also called Schottky noise, quantum noise, and shot noise. To model it in a semiconductor junction, we can use the formula given by equation [1.5]:

$$\overline{|\iota_n|^2} = 2q \cdot I_0 \cdot df \qquad [1.5]$$

where:

– q is the electron's charge;

– I_0 is the junction's bias current.

This shot noise is a white noise, that is to say a spectral power density independent of the frequency, provided that the frequency considered is less than the transit time of the carriers in the junction.

Although different in nature from thermal noise, we can model the shot noise by defining an equivalent resistance R_{neq} for shot noise and using it in equation [1.5] as $\overline{|\iota_n|^2} = 4 \cdot kT \cdot R_{neq} \cdot df$. We can show that this equivalent resistance R_{neq} is equal to half R_{Dyn}, the dynamic resistance of the junction around its point of polarization with:

$$R_{Dyn} = \frac{kT}{qI_0}$$

1.2.4. *Generation / recombination noise*

In a semiconductor, the current is related to the number of carriers and these carriers come from the ionized dopants. These ionized dopants have a tendency for donors to try to recover the electron they have given away; it is the process of recombination. For acceptors, the same recombination phenomenon exists with the holes. These recombination phenomena are randomly distributed and the recombined carriers no longer participate in the current.

Since these recombinant carriers are not necessary for the stability of the crystalline structure of the semiconductor material, they are released after a certain time by those that captured them. It is the phenomenon of generation which, too, is randomly distributed.

This generation / recombination phenomenon, by randomly varying the number of carriers capable of moving in the semiconductor, is therefore responsible for generating / recombining current noise. This noise is rather a low frequency one, given the time constants involved, and it is often drowned in excess noise.

1.2.5. *Excess noise*

Excess noise is often qualified as noise with poorly known origins but nevertheless, it can be said that this noise does not have an intrinsic origin to the semiconductor material. It is rather related to the constraints imposed for the realization of electronic components, among which we can quote:

– the imperfections of the crystalline structure which result in dislocations;

– the unintentional impurities that remain in the semiconductor materials, despite the purification techniques used. These impurities are responsible for trapping / de-trapping phenomena similar to the generation / recombination phenomena;

– the finite dimensions of the components and the layers used which are of different natures and which generate interface phenomena.

All these phenomena have in common to generate noise at low frequency which tends to see their power spectral densities decrease when the frequency increases. This is called noise in $1/f$ or in $1/f^2$. This type of noise tends to disappear in thermal noise beyond a certain frequency, as shown in Figure 1.6.

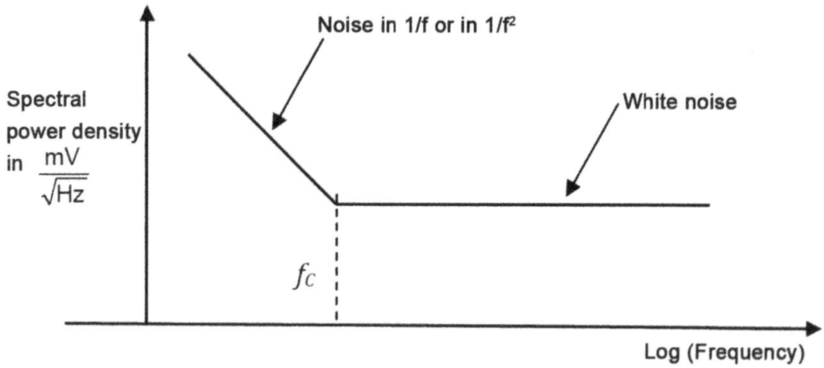

Figure 1.6. *Spectral power density: excess noise and thermal noise*

The cutoff frequency f_C depends on the semiconductor material in question and is of the order of 10 kHz for silicon and 100 MHz for gallium arsenide. In general, this excess noise is not a problem in radiofrequency or microwave frequencies because the frequencies involved are greater than these cutoff frequencies. Nevertheless, when designing radio-frequency or microwave oscillators, it is necessary to take into account the excess noise that is responsible for the phase noise of these oscillators.

1.3. Noise factor

1.3.1. *Definition*

In analog signal processing, we are very often in the canonical situation of Figure 1.7, where a signal source is connected to the input of a signal processing two-port whose output is connected to a signal receiver. The signal source, which can be seen as a voltage generator consisting of an ideal voltage source E in series with an internal impedance Z_S, also delivers noise; at a minimum, the thermal noise due to Z_S. The processing two-port will perform its processing on the signal delivered by the source and treat in the same way the noise delivered by the source. In addition, as the two-port contains electronic devices that are all noisy, it will add its own noise to the output signal. Intuitively, one might think that the signal-to-noise ratio at the output is smaller than the signal-to-noise ratio at the input, because of the noise added by the two-port. By definition, the noise factor F of a two-port provides information on the degradation of the signal-to-noise ratio between the input and the output of a two-port, as indicated by equation [1.6]. This quantity is always greater than or equal to 1.

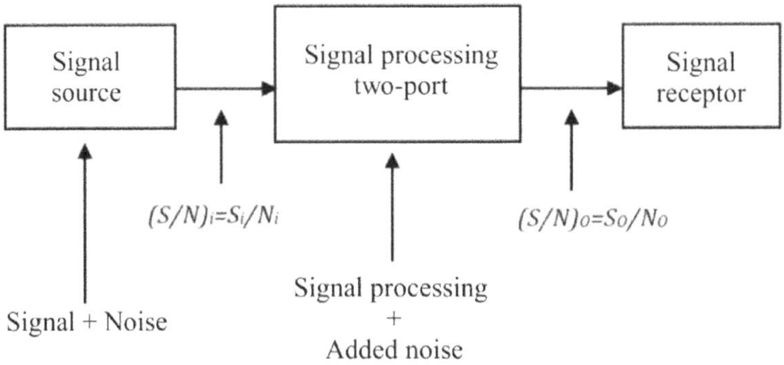

Figure 1.7. *Signal processing and signal-to-noise ratio*

$$F = \left(S_i/N_i\right) / \left(S_O/N_O\right) \quad [1.6]$$

Assuming that the signal power S_O, measured at the output, is equal to $G.S_i$ where G is the power gain of the two-port, then it can be written that the noise power, measured at the output, is equal to $N_O = N_A + G \cdot N_i$ where N_A denotes the noise power added by the two-port, measured at the output. This noise of the two-port is not correlated with that of the source and therefore, in power, is added to $G \cdot N_i$.

Under these conditions, the equation [1.6] becomes equation [1.7] where it is clear that F is always greater than or equal to 1:

$$F = \frac{S_i \cdot N_O}{S_O \cdot N_i} = \frac{N_A + G \cdot N_i}{G \cdot N_i} = 1 + \frac{N_A}{G \cdot N_i} \quad [1.7]$$

Often, when talking about power, it is convenient to use dB. The expression [1.7] is that of the linear noise factor. To get it in dB, use the expression [1.8]:

$$F_{dB} = 10 \cdot log_{10}(F_{lin}) = NF \quad [1.8]$$

We must be careful because often in French we use – wrongly – the same name for F_{dB} and F_{lin}. English speakers are more cautious and use *noise figure* for F_{dB} and *noise factor* for F_{lin}.

Here, the title of section 1.3 is "adequate" because the expressions of equations [1.6] and [1.7] are those of the noise factor; on the other hand, very few French-speaking scientists in the field use the term "noise figure" for F_{dB}.

As the problem arises just when one has to use a formula (in a datasheet, there is no ambiguity because the unit is indicated like this: nothing for the noise factor or dB for the noise figure), some authors write F for the noise factor in linear and NF for its value in dB.

1.3.2. *Reference temperature for the noise factor*

The expression of equation [1.7] involves N_i the noise power of the source in the calculation of the noise factor of the two-port to be characterized in degradation of the signal / noise ratio. The noise factor is therefore not an intrinsic feature of the two-port because it involves the noise power of the source connected to the input of this two-port: it is therefore a relative measure of noise behavior of the two-port. To remedy this problem, one solution has been proposed: to use a single reference temperature when using the concept of noise factor. This reference temperature is $T_0 = 290\ K$.

Another solution is to use an intrinsic quantity for the two-port to define its noise behavior; it is the notion of equivalent noise temperature. Taking the equation [1.7] and assuming that the source is at a temperature T_i, then the noise power delivered by this source is $N_i = k \cdot T_i \cdot df$. The quantity $\frac{N_A}{G}$ represents the noise power added by the two-port, but measured at its input. If we model this noise power as thermal noise, we can write equation [1.9] and use T_e as the equivalent noise temperature of the two-port:

$$\frac{N_A}{G} = k \cdot T_e \cdot df \quad [1.9]$$

This equivalent noise temperature T_e is then a quantity which concerns only the for-port to be qualified as noise through N_A and G; except that the quantity $\frac{N_A}{G}$ depends on the load presented at the input of the four-port, but this is also the case when the notion of noise factor F is used.

This equivalent noise temperature T_e reflects the fact that if the quadripole is used to process the signal from a source at the temperature T_S, everything happens as if the source was at $T_S + T_e$ then processed with the same two-port, but this time considered as noiseless.

$$F = 1 + \frac{T_e}{T_0} \text{ or } T_e = T_0 \cdot (F\text{-}1) \quad [1.10]$$

The relationship between noise factor and equivalent noise temperature is given by equation [1.10]. In the rest of the book, we will be interested only in the noise factor.

1.3.3. *Importance of the noise factor in telecommunications*

Figure 1.8 shows the simplified block diagram of the receiving part of a digital radio transmission system. The received signal has a signal-to-noise ratio S_i/N_i. After low noise amplification, down-frequency conversion and demodulation; the signal has a signal-to-noise ratio S_o/N_o which is lower than S_i/N_i because the three stages traveled each add a contribution in noise.

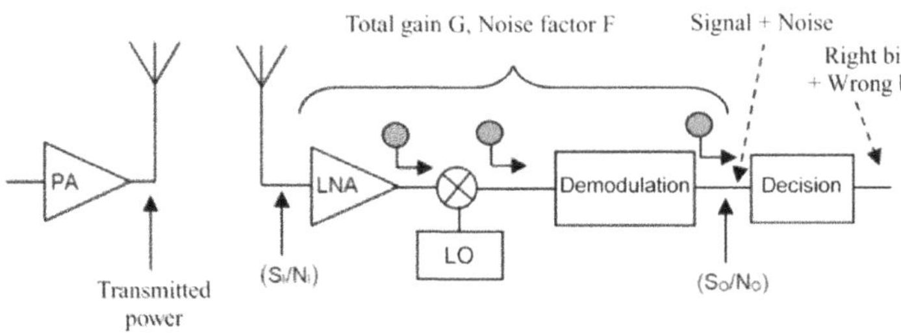

Figure 1.8. *Synoptic of the receiver of a digital wireless transmission*

The decision system then samples the signal and assigns it a numerical value according to criteria which are thresholds making it possible to distinguish the different symbols emitted as a function of the digital values to be transmitted. The power noise N_o randomly modifies the values of the samples corresponding to the emitted symbols. In some cases, the noise level will cause the observed sample to pass through a domain that does not match the transmitted symbol and a transmission error occurs. For Gaussian additive noise, depending on the signal-to-noise ratio at the input of the decision system, the type of modulation used and the bit rate, the BER, bit error rate, can be calculated; the BER being equal to the number of false bits received over the total number of bits transmitted.

For example, for QPSK modulation (Quadrature phase-shift keying processing of bits 2 by 2), for the most favorable bandwidth conditions and for a given signal/noise ratio at the input of the receiver S_i/N_i=16 dB, we can calculate, the corresponding BER given in Table 1.1.

NF	*BER*
3 dB	4×10^{-6}
4 dB	3×10^{-5}
5 dB	2×10^{-4}

Table 1.1. *BER as a function of NF for a QAM4 with* S_i/N_i=*16 dB*

We see that the bit error rate is multiplied by 50 when we go from *NF* = 3 dB to NF = 5 dB. The advantage of an optimization of the noise factor of the receiver is thus immediately understood and therefore of being able to calculate a noise factor and to measure this quantity.

1.4. Noise in two-ports

1.4.1. *Representation of noise in two-ports*

The general principle for representing noise in a two-port placed at temperature T is to use two noise generators outside the two-port (Rothe *et al.* 1956) and to consider that the two-port is free of any noise; that is to say, it behaves as if it were placed at the fictional temperature 0 K.

With regard to generators, we have several choices:

– two noise voltage generators e_1 and e_2 placed in series on its two ports, as shown in Figure 1.9;

– two noise current generators i_1 and i_2 placed in parallel on its two accesses, as indicated in Figure 1.10;

– two noise generators placed only on the input, one in voltage e in series and one in current i in parallel, as indicated in Figure 1.11;

– two noise power wave generators b_{n1} and b_{n2} placed on each of its two accesses which are described in Appendix 5.

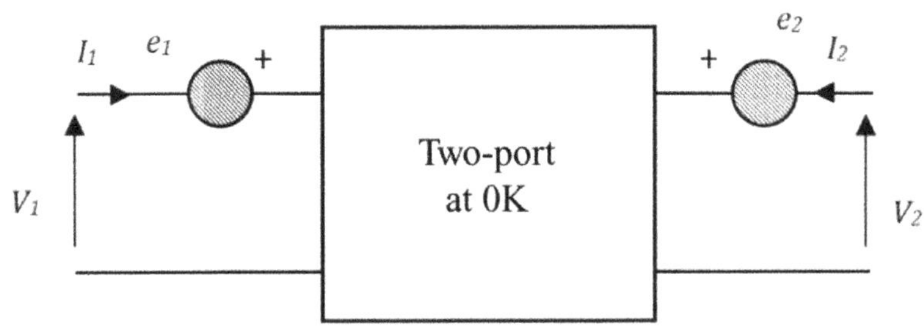

Figure 1.9. *Noise representation by two noise voltage generators*

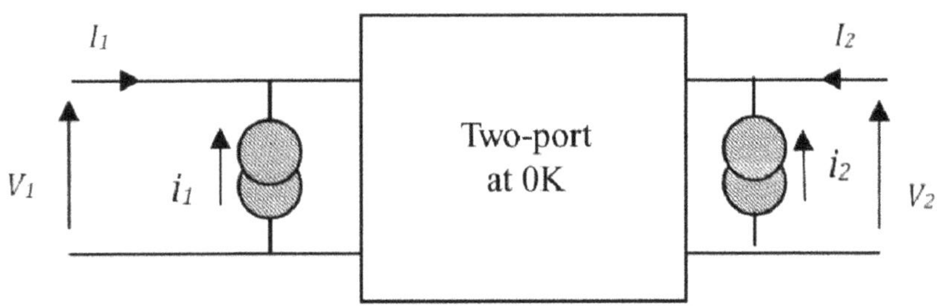

Figure 1.10. *Noise representation by two noise current generators*

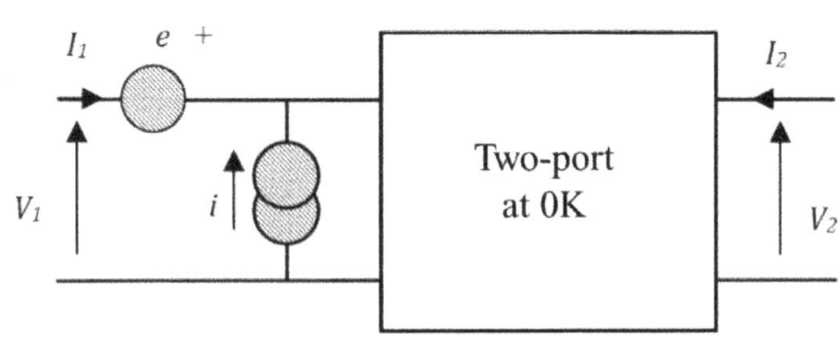

Figure 1.11. *Noise representation by a noise current generator and a noise voltage generator*

For these four representations, it should be understood that the two generators are equivalent generators, in the sense that it is possible to calculate them using Thévenin and Norton's theorems, for all the phenomena responsible for noises in the two-port and their associated generator; they do not necessarily have a physical reality that can be explained immediately. In the same way, as they result from the superposition of physical phenomena internal to the two-port, but only observed on the accesses of this two-port, they are often partially correlated.

Of these four main representations, the representation of Figure 1.11 is the most used because all the noise is brought back to the input of the two-port and this facilitates, as we will see later, the calculation of the various quantities. It is with this representation that we will continue to explore the notion of noise factor using the Rothe notation (Rothe *et al.* 1956).

1.4.2. *Expression of the noise factor of a two-port*

This calculation is done using the representation of the noise in the quadrupole of Figure 1.11 for which the input of the two-port is closed on a source admittance Y_S itself modeled in noise by a noise current generator whose average quadratic value squared is given by the Nyquist formula, i.e. $|\overline{i_{nS}}|^2 = 4 \cdot kT_0 \cdot G_S df$ where G_S is the real part of Y_S and T_0 is the reference temperature 290 K. The corresponding circuit is given in Figure 1.12.

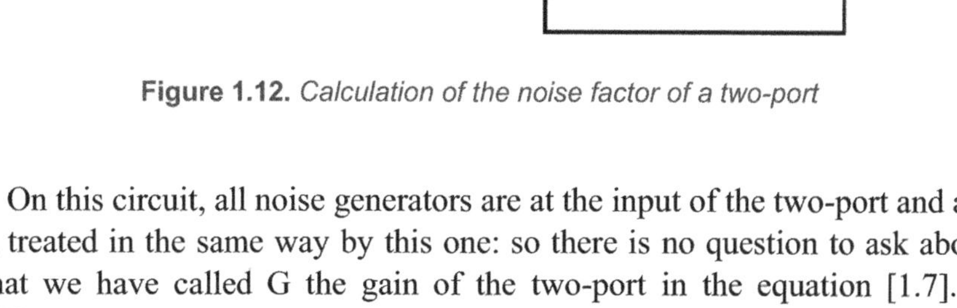

Figure 1.12. *Calculation of the noise factor of a two-port*

On this circuit, all noise generators are at the input of the two-port and are all treated in the same way by this one: so there is no question to ask about what we have called G the gain of the two-port in the equation [1.7]. It suffices to calculate the current generator equivalent to these three noise generators and to calculate two noise powers at the input of the two-port, and this in any admittance, which may be Y_E the input admittance of the two-port if you want to be reassured:

– that due to the source admittance which corresponds to the term N_i of equation [1.7], that is to say when $e = i = 0$;

– that due to the quadripole which corresponds to the term $\frac{N_A}{G}$ of equation [1.7], that is to say when $i_{nS} = 0$.

Under these conditions, the two-port noise factor is given by the expression in equation [1.11]:

$$F = 1 + \frac{|\overline{e \cdot Y_S + i}|^2}{|\overline{i_{nS}}|^2} = F(Y_S) \qquad [1.11]$$

So, we see that it is the noise factor of the two-port, but it also depends on the choice of the source admittance Y_S. It is therefore better to denote it as $F(Y_S)$.

1.4.3. *Minimum noise factor of a two-port*

Since the noise factor of a two-port depends on the source admittance chosen, it is legitimate to ask the following question: for a given four-port, what is the value of Y_S which gives the lowest noise factor. To answer this question, we will take the notation used in (Rothe *et al.* 1956) and given by equations [1.12a] to [1.12e]:

$$Y_S = G_S + jB_S \qquad [1.12a]$$

$$|\bar{e}|^2 = 4.kT_0.R_N df \qquad [1.12b]$$

$$i = i_n + Y_{Cor}.e \qquad [1.12c]$$

$$Y_{Cor} = G_{Cor} + jB_{Cor} \qquad [1.12d]$$

$$|\bar{i_n}|^2 = 4.kT_0.G_n df \qquad [1.12e]$$

Equation [1.12a] simply translates that admittance is a priori a complex quantity.

Equation [1.12b] defines R_N a quantity called equivalent noise resistance at the input of the two-port which makes it possible to express simply $|\bar{e}|^2$ at the reference temperature T_0. In the general case, this resistance R_N has no immediate physical reality.

Equation [1.12c] reflects the fact that i and e are equivalent generators seen at the input of the two-port and that they can be partially correlated according to how the noise is produced in the two-port. They are uncorrelated if Y_{Cor} is zero and fully correlated if i_n is zero. Therefore i_n is the uncorrelated part to e of i. Since i is a current and e a voltage, the correlation coefficient from i to e is well homogeneous with admittance.

Equation [1.12d] expresses the fact that the correlation admittance between i and e is a priori a complex quantity.

Equation [1.12e] defines G_n as a quantity called noise equivalent conductance at the input of the two-port which allows us to simply express $|\bar{i_n}|^2$ at the reference temperature T_0. In the general case, this conductance G_n has no immediate physical reality. Be careful, there is no relation between G_n and R_N (especially not $G_n = 1/R_N$), hence the indices N and n to distinguish these two quantities.

Using the notations given by equations [1.12a] to [1.12e], we can rewrite the two-port noise factor for a source admittance Y_S, as indicated in equation [1.13]:

$$F(Y_S) = 1 + \frac{\left|\overline{e\cdot(Y_S+Y_{Cor})+i_n}\right|^2}{|\overline{i_{nS}}|^2} = 1 + \frac{R_N\cdot|Y_S+Y_{Cor}|^2+G_n}{G_S} \quad [1.13]$$

From equation [1.13], it suffices to derive $F(Y_S)$ with respect to B_S and with respect to G_S then to cancel these two derivatives to find F_{Min} the minimum noise factor and Y_{Sopt} the optimal source admittance, for the noise, of the two-port which gives F_{Min}.

The derivative of $F(Y_S)$ with respect to B_S is written as $\frac{2\cdot R_N(B_S+B_{Cor})}{G_S}$ and it is necessary to let $B_s = -B_{Cor}$ to cancel this derivative.

For $B_s = -B_{Cor}$, the numerator of the derivative of $F(Y_S)$ with respect to G_S is written as $G_S^2 - (G_{Cor}^2 + \frac{G_n}{R_N})$, and it is necessary to let $G_S = \sqrt{G_{Cor}^2 + \frac{G_n}{R_N}}$ to cancel this derivative.

We can thus deduce Y_{Sopt} the optimal source admittance for the noise whose expression is given by the equation [1.14]:

$$Y_{Sopt} = G_{Sopt} + jB_{Sopt} = \sqrt{G_{Cor}^2 + \frac{G_n}{R_N}} - jB_{Cor} \qquad [1.14]$$

For this optimal source admittance for noise, the minimum noise factor is given by equation [1.15]:

$$F_{Min} = 1 + 2 \cdot R_N \cdot \left(G_{Sopt} + G_{Cor}\right) = 1 + F_{eMin} \qquad [1.15]$$

The quantity F_{eMin} is the minimum excess noise factor of the two-port.

By using the equations [1.14] and [1.15] in [1.13], we obtain the expression given by the equation [1.16] which gives the noise factor for any admittance:

$$F(Y_S) = F_{Min} + \frac{R_N \cdot |Y_S - Y_{Sopt}|^2}{G_S} \qquad [1.16]$$

Since the noise factor of a two-port is a quantity that is used in radiofrequency and microwave frequencies, it may be useful to express it as a function of Γ_S the reflection coefficient of the source, as detailed in equation [1.17]:

$$F(\Gamma_S) = F_{Min} + \frac{4 \cdot R_N}{R_0} \frac{|\Gamma_S - \Gamma_{Sopt}|^2}{|1 + \Gamma_{Sopt}|^2 \cdot (1 - |\Gamma_S|^2)} \qquad [1.17a]$$

$$\Gamma_S = \frac{Z_S - R_0}{Z_S + R_0} = \frac{G_0 - G_S}{G_0 + G_S} \qquad [1.17b]$$

$$\Gamma_{Sopt} = \frac{Z_{Sopt} - R_0}{Z_{Sopt} + R_0} = \frac{G_0 - Y_{Sopt}}{G_0 + Y_{Sopt}} \qquad [1.17c]$$

Where Γ_{Sopt} is the optimum source reflection coefficient for noise.

1.4.4. *Inverse relations*

Implicitly, in the previous section, we have seen the four quantities $|\bar{e}|^2, |\bar{\iota_n}|^2$ and $Y_{Cor} = G_{Cor} + jB_{Cor}$ and we have deduced the four noise parameters used to describe the behavior in noise of a two-port that are F_{Min}, R_N, $|\Gamma_{Sopt}|$ and $Arg(\Gamma_{Sopt})$. We will now do the opposite; that is to say from F_{Min}, R_N and Γ_{Sopt}, quantities which are accessible to the measure, as we will see in Chapter 5, and express $|\bar{e}|^2, |\bar{\iota_n}|^2$ and $Y_{Cor} = G_{Cor} + jB_{Cor}$ that can be used to describe the noise of the two-port when it is used in a circuit with other components.

The knowledge of R_N makes it possible to write directly the equation [1.18]:

$$e = 2\sqrt{kT_0 \cdot df \cdot R_N} \qquad [1.18]$$

The knowledge of Γ_{Sopt} makes it possible to write the equality [1.19] which gives the real part and the imaginary part of Y_{Sopt}. The operator *Im* is the "imaginary part" operator on a complex:

$$Y_{Sopt} = G_0 \cdot \frac{1-\Gamma_{Sopt}}{1+\Gamma_{Sopt}} = G_0 \cdot \frac{1-|\Gamma_{Sopt}|^2}{|1+\Gamma_{Sopt}|^2} - G_0 \cdot \frac{2j.Im(\Gamma_{Sopt})}{|1+\Gamma_{Sopt}|^2} = G_{Sopt} + jB_{Sopt} \qquad [1.19]$$

We can then express Y_{Cor} by using $B_{Cor} = -B_{Sopt}$ and $G_{Cor} = \frac{F_{eMin}}{2R_N} - G_{Sopt}$ as given by equations [1.20]:

$$B_{Cor} = G_0 \cdot \frac{2j.Im(\Gamma_{Sopt})}{|1+\Gamma_{Sopt}|^2} \qquad [1.20a]$$

$$G_{Cor} = \frac{F_{eMin}}{2R_N} - G_0 \cdot \frac{1-|\Gamma_{Sopt}|^2}{|1+\Gamma_{Sopt}|^2} \qquad [1.20b]$$

It only remains to obtain G_n using [1.14] as written in equation [1.21]:

$$G_n = R_N \cdot \left(G_{Sopt}{}^2 - G_{Cor}{}^2\right) = \frac{F_{eMin}}{2} \cdot \left(2G_0 \cdot \frac{1-|\Gamma_{Sopt}|^2}{|1+\Gamma_{Sopt}|^2} - \frac{F_{eMin}}{2R_N}\right) =$$

$$F_{eMin} \cdot G_0 \cdot \frac{1-|\Gamma_{Sopt}|^2}{|1+\Gamma_{Sopt}|^2} - \frac{F_{eMin}^2}{4R_N} \qquad [1.21]$$

1.5. Characterization of noise in a two-port

Thanks to equations [1.16] and [1.17], we find that to know the noise behavior of a two-port, we need to define or measure four quantities which are:

– F_{Min} the minimum noise factor or $T_{eMin} = T_0 \cdot (F_{Min} - 1)$ the minimum noise equivalent temperature;

– R_N the equivalent noise resistance at the input;

– Y_{Sopt} the optimal source admittance for noise that contains two independent parts of information G_{Sopt} and B_{Sopt} or Γ_{Sopt} which also contains two independent pieces of information $|\Gamma_{Sopt}|$ and $Arg(\Gamma_{Sopt})$.

To arrive at determining these four quantities, it is necessary to have a noise measurement bench which makes it possible to vary the source admittance presented at the input of the two-port. The synoptic of such a bench is given in Figure 1.13.

In such a measurement bank, we can:

– vary the reflection coefficient Γ_S presented to the object under test and thus to measure the noise factor of the DUT (device under test) for these different values by setting the switches SW1 and SW2 in the red position and measuring the power of noise with a spectrum analyzer (SA) preceded by a low noise amplifier (LNA). The method will be explained in more detail in Chapter 5;

– measure with the network analyzer the S parameters of the DUT which will be needed to, inter alia, determine the gain of the DUT;

– calibrate the bench in noise and tuner by connecting P1 and P2 directly.

The bias tees are there to allow biasing the DUT, if it does not have its own bias circuits.

Such a measurement bench can only be used if it is automated thanks to a computer controlling the vectorial network analyzer (VNA), the switches, the spectrum analyzer and the noise source.

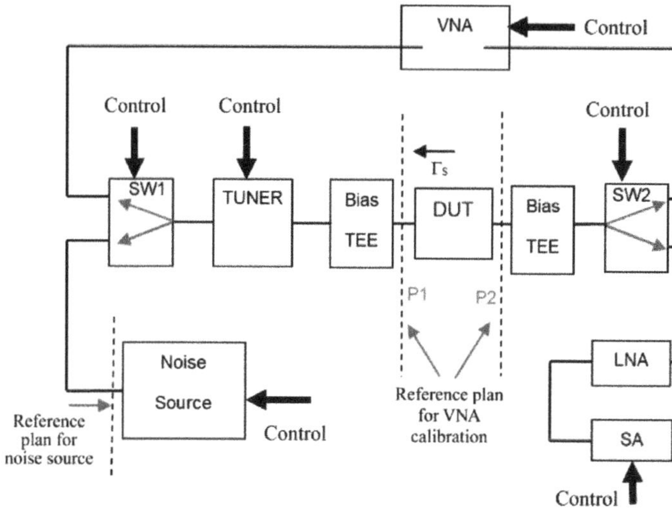

Figure 1.13. *Noise characterization bench of a two-port. For a color version of this figure, see www.iste.co.uk/fouquet/electronics.zip*

Figure 1.14 shows the characterization bench that was installed at ESIGELEC as part of a research project. Its detailed operation is explained in Chapter 5.

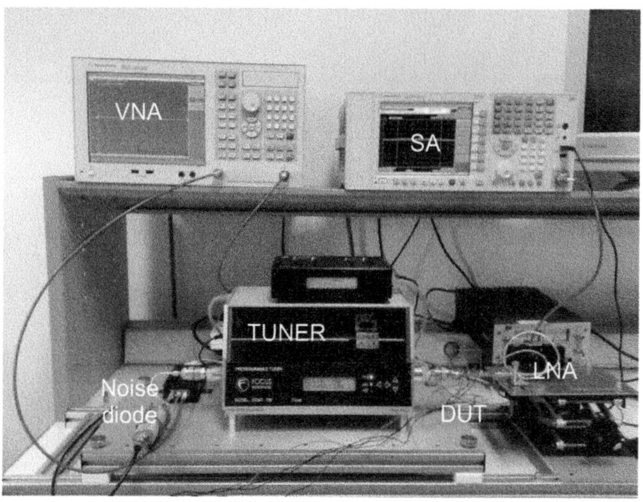

Figure 1.14. *ESIGELEC noise characterization bench. For a color version of this figure, see www.iste.co.uk/fouquet/electronics.zip*

1.6. Conclusion

This first chapter allowed us to see the origins of spontaneous fluctuations in electronic components responsible for background noise in electronic circuits. We have also defined the notion of mean square value for a current noise source or a voltage noise source. For the two-ports, we used two equivalent sources of noise and highlighted the correlation that can exist between these two sources. From there, we have highlighted the fact that a two-port can be completely described with respect to noise by four quantities which are the following:

– F_{Min} the minimum noise factor or $T_{eMin} = T_0.(F_{Min}-1)$ the minimum equivalent noise temperature;

– R_N the equivalent noise resistance at the input of the quadripole;

– Y_{Sopt} the optimal source admittance for noise that contains two independent pieces of information G_{Sopt} and B_{Sopt} or Γ_{Sopt}which also contains two independent pieces of information $|\Gamma_{Sopt}|$ and $Arg(\Gamma_{Sopt})$.

We now have the necessary tools (equivalent generators and noise calculation method) to address more complex topics, such as cascading noisy two-ports and noise factor measurement.

2

Friis Formula

2.1. Introduction

The objective of this chapter is to explicitly demonstrate the Friis formula, which is often written abusively, for two two-ports Q1 and Q2 in cascade, in the form given by the equation [2.1]. What is not explicitly stated in equation [2.1] is that:

– the noise factor of the Q1 Q2 association in cascade, denoted F_{12}, is relative to the source admittance seen by the first two-port Q1, i.e. $F_{12}(Y_S)$;

– the noise factor, denoted F_1, is that of two-port Q1, and is relative to the source admittance seen by this first two-port Q1 which is Y_S, i.e. $F_1(Y_S)$;

– the noise factor, denoted F_2, is that of two-port Q2 and is relative to the source admittance seen by this two-port Q2 which is Y_{OUT1}, i.e. F_2 (Y_{OUT1}); Y_{OUT1} being the output admittance of Q1 when its input is loaded by Y_S;

– the power gain, denoted G_1, is in fact the available power gain of the two-port Q1 and that it is therefore the transducer power gain computed for a source admittance Y_S and for a load admittance $Y_L = Y_{OUT1}{}^{*}$

The "true" Friis formula (Friis 1944) must therefore be written as indicated by equation [2.2]:

$$F_{12} = F_1 + (F_2 - 1)/G_1 \qquad [2.1]$$

$$F_{12}(Y_S) = F_1(Y_S) + \frac{F_2(Y_{OUT1}) - 1}{G_{Avai1}(Y_S)} \qquad [2.2]$$

The form given by equation [2.1] is acceptable only in a very particular case which corresponds to the calculation of the noise factor of two two-ports Q1 and Q2 in cascade under the following conditions:

– we want to calculate $F_{12}(Y_S)$ for $Y_S = G_0 = 1/R_0$ where R_0 is the reference impedance 50 Ω;

– the two-ports Q1 and Q2 are adapted, at the input and at the output, on $R_0 = 50$ Ω;

It is all the more important to understand this Friis formula – and the quantities that intervene in it – that its use, well understood, is essential to make noise factor measurements on a two-port as we will see in Chapter 4.

To demonstrate rigorously the Friis formula written in equation [2.2], we will use the method described in section 2.2.

2.2. Calculation method

To represent the noise in a noisy two-port, we know that we can use two partially correlated equivalent noise sources, one in voltage and the other in current, placed at the input of the two-port as shown in Figure 2.1. Under these conditions (Rothe *et al.* 1956), the noise factor of this two-port $F(Y_S)$, when its input is terminated with an admittance source Y_S, is given by equation [2.3] taking into account the arbitrary directions chosen for e and i:

$$F(Y_S)_{lin} = 1 + \frac{\overline{|e \cdot Y_S + i|^2}}{\overline{|i_S|^2}} \qquad [2.3]$$

With i_S being the source of the noise current of the source admittance Y_S; or $\overline{|i_S|^2} = 4kTdf \cdot G_S$, with G_S being the real part of Y_S. The writing $F(Y_S)_{lin}$ is here to specify that all the equations present in this chapter concern noise factors expressed in linear and not in dB.

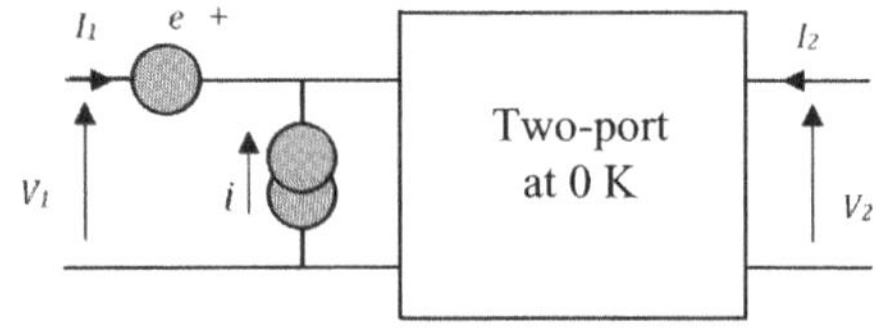

Figure 2.1. *Two-port at 0K with voltage sources and equivalent noise current*

As stated in Chapter 1, equation [2.3] can be established by calculating the two noise powers $N|_{e=i=0}$ (due to i_S alone) and $N|_{iS=0}$ (due to e and i) collected in the input admittance of the two-port and by writing $F=1+ N|_{iS=0} /N|_{e=i=0}$ since i_S is not correlated with e nor with i.

The shape of equation [2.3] is the main interest of the noise representation of Figure 2.1, where all the noise due to the two-port is brought back to its input.

With this same representation of Figure 2.1, we can also write the matrix form given by equation [2.4] where the Y_{ij} are the admittance parameters of the two-port. This matrix notation, seldom used although very practical in noise calculation, can be very easily established by calculating the contributions of e and i to I_1 and I_2 when $V_1=V_2=0$:

$$\begin{vmatrix} I_1 \\ I_2 \end{vmatrix} = \begin{bmatrix} Y_{11} & Y_{12} \\ Y_{21} & Y_{22} \end{bmatrix} \cdot \begin{vmatrix} V_1 \\ V_2 \end{vmatrix} + \begin{bmatrix} Y_{11} & -1 \\ Y_{21} & 0 \end{bmatrix} \cdot \begin{vmatrix} e \\ i \end{vmatrix} \quad [2.4]$$

To establish the Friis formula, we will consider two two-ports Q1 and Q2 in cascade and for which we know the admittance matrices noted $[Y_1]$ for Q1 and $[Y_2]$ for Q2 and the equivalent noise sources e_1, i_1 for Q1 and e_2, i_2 for Q2 as shown in Figure 2.2.

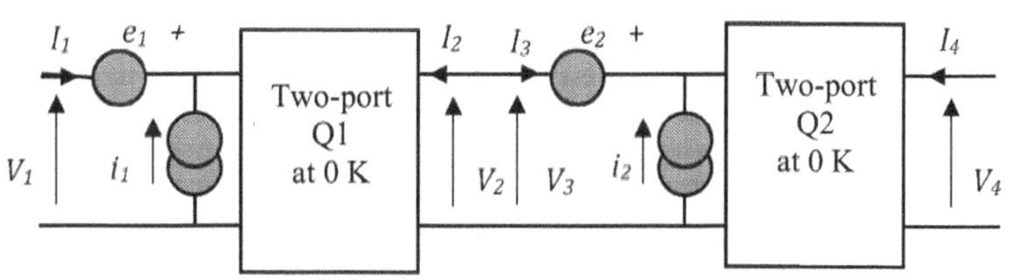

Figure 2.2. *Two-ports Q1 and Q2 in cascade at 0K with voltage sources and equivalent noise current*

This arrangement of Figure 2.2 can be described by an admittance matrix $[Y_{Tot}]$ and two noise sources e_{Tot} and i_{Tot} as indicated in equation [2.5]:

$$\begin{vmatrix} I_1 \\ I_4 \end{vmatrix} = \begin{bmatrix} Y_{11Tot} & Y_{12Tot} \\ Y_{21Tot} & Y_{22Tot} \end{bmatrix} \cdot \begin{vmatrix} V_1 \\ V_4 \end{vmatrix} + \begin{bmatrix} Y_{11Tot} & -1 \\ Y_{21Tot} & 0 \end{bmatrix} \cdot \begin{vmatrix} e_{Tot} \\ i_{tot} \end{vmatrix} \quad [2.5]$$

To be able to calculate the noise factor of this association Q1, Q2 in cascade, it is enough to:

– calculate Y_{11Tot} and Y_{21Tot};

– express the contributions of the noise generators e_1, i_1, e_2 and i_2 to I_1 and I_4;

– identify e_{Tot} and i_{Tot};

– calculate $F_{12}(Y_S)$ using equation [2.3].

To establish the Friis formula, we will only need to identify the different components of $F_{12}(Y_S)$ to get $F_1(Y_S)$, $F_2(Y_{OUT1})$ and $G_{Avai1}(Y_S)$.

2.3. Calculation of the admittance parameters of the Q1, Q2 association in cascade

We must calculate Y_{11Tot} and Y_{21Tot}. For this, we use the equations given by [2.6] with $V_4 = 0$:

$$\begin{vmatrix} I_1 \\ I_2 \end{vmatrix} = \begin{bmatrix} Y_{11} & Y_{12} \\ Y_{21} & Y_{22} \end{bmatrix} \cdot \begin{vmatrix} V_1 \\ V_2 \end{vmatrix} \quad [2.6a]$$

$$\begin{vmatrix} I_3 \\ I_4 \end{vmatrix} = \begin{bmatrix} Y_{33} & Y_{34} \\ Y_{43} & Y_{44} \end{bmatrix} \cdot \begin{vmatrix} V_3 \\ V_4 \end{vmatrix} \quad [2.6b]$$

$$I_2 = -I_3 \quad [2.6c]$$

$$V_2 = V_3 \quad [2.6d]$$

By making substitutions in [2.6] with $V_4 = 0$, we obtain the expressions given by equations [2.7] and [2.8]:

$$Y_{11Tot} = Y_{11} - \frac{Y_{21}Y_{12}}{Y_{22}+Y_{33}} \quad [2.7]$$

$$Y_{21Tot} = -\frac{Y_{21}Y_{43}}{Y_{22}+Y_{33}} \quad [2.8]$$

These equations can also be obtained by using the flow graph and Mason's rule. The reader can refer to Appendix 3 for more details on the use of these two tools.

Although this is not necessary for the rest of the proof, we can calculate Y_{22Tot} and Y_{12Tot} with equation [2.6] by having $V_1 = 0$ or by using again the flow graph and Mason's rule. We obtain the expressions given by equations [2.9] and [2.10]:

$$Y_{22Tot} = Y_{22} - \frac{Y_{21}Y_{12}}{Y_{22}+Y_{33}} \quad [2.9]$$

$$Y_{12Tot} = -\frac{Y_{12}Y_{34}}{Y_{22}+Y_{33}} \quad [2.10]$$

2.4. Contribution of noise generators e_1, i_1, e_2 and i_2 to I_1 and I_4

2.4.1. *Introduction*

This computation is done for $V_1 = V_4 = 0$ by considering successively the contributions of the four noise generators e_1, i_1, e_2 and i_2 to currents I_1 and I_4. This method, although repetitive, is much simpler than tackling the problem by directly taking into account the four generators. The superposition theorem that is implicitly used applies to noise generators as long as we do not use the root mean square values. To switch to the root mean square values, we must take into account the correlation that can exist between generators.

2.4.2. *Contribution of the noise generator e_1*

By replacing Q1 and Q2 by their equivalent diagram and taking into account only the noise generator e_1, we obtain the circuit of Figure 2.3.

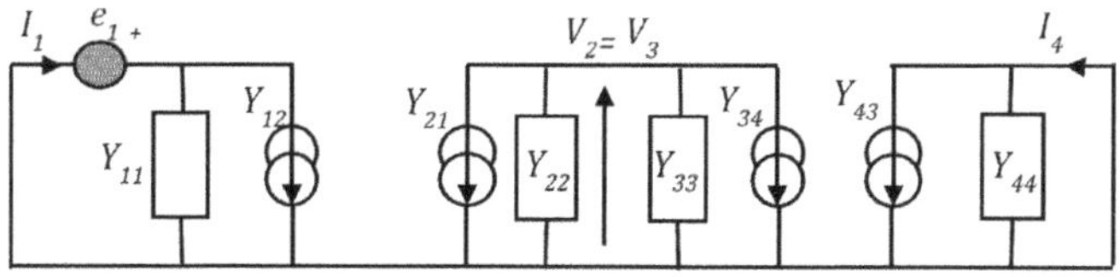

Figure 2.3. *Equivalent diagram of Q1 and Q2 in cascade: contribution from generator e_1 to I_1 and I_4*

In this equivalent circuit, it must be taken into account that V4 = 0 and that the current of the controlled generator $Y_{34}V_4$ is zero as well as the current in Y_{44}. On the other hand, the control voltage of the controlled current generator Y_{21} is e_1 and the control voltage of the current generator Y_{43} and Y_{12} is:

$$V_2=V_3=-Y_{21}e_1/(Y_{22}+Y_{33})$$

We can then write equations [2.11] and [2.12]:

$$I_1 = \left(Y_{11} - \frac{Y_{21}Y_{12}}{Y_{22}+Y_{33}}\right) \cdot e_1 = Y_{11Tot} \cdot e_1 \quad [2.11]$$

$$I_4 = -\frac{Y_{21}Y_{43}}{Y_{22}+Y_{33}} \cdot e_1 = Y_{21Tot} \cdot e_1 \quad [2.12]$$

2.4.3. *Contribution of the noise generator i_1*

By replacing Q1 and Q2 by their equivalent circuit and taking into account only the noise generator i_1, we obtain the circuit of Figure 2.4. In this arrangement, the control voltage of the controlled current generator Y_{21} is zero and no current flows into Y_{11}.

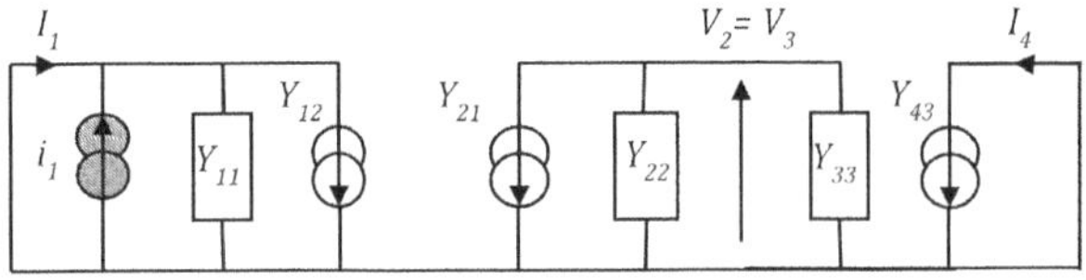

Figure 2.4. *Equivalent diagram of Q1 and Q2 in cascade: contribution of the generator i_1 to I_1 and I_4*

We can therefore write equations [2.13] and [2.14]:

$$I_1 = -i_1 \quad [2.13]$$

$$I_4 = 0 \quad [2.14]$$

2.4.4. *Contribution of the noise generator e_2*

By replacing Q1 and Q2 by their equivalent circuit and taking into account only the noise generator e_2, we obtain the circuit of Figure 2.5. In this arrangement, the control voltage of the controlled current generator Y_{21} is zero and no current flows into Y_{11}. The control voltage of the controlled current generator Y_{12} is $-Y_{33}{\cdot}e_2/(Y_{22}+Y_{33})$. The control voltage of the controlled current generator Y_{43} is $Y_{22}{\cdot}e_2/(Y_{22}+Y_{33})$. We can therefore write equations [2.15] and [2.16]:

$$I_1 = -\left(\frac{Y_{33}Y_{12}}{Y_{22}+Y_{33}}\right) \cdot e_2 \quad [2.15]$$

$$I_4 = \frac{Y_{22}Y_{43}}{Y_{22}+Y_{33}} \cdot e_2 \quad [2.16]$$

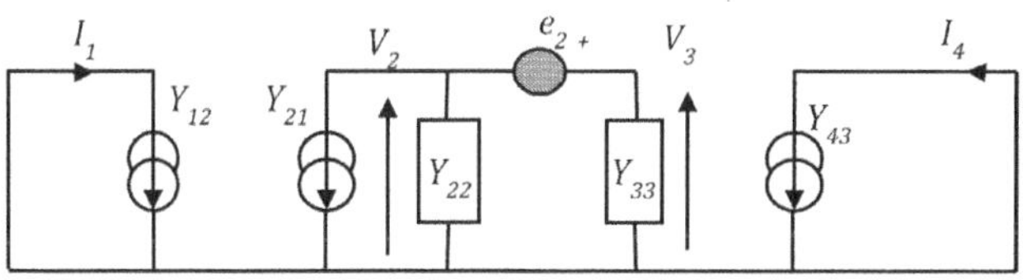

Figure 2.5. *Equivalent diagram of Q1 and Q2 in cascade: contribution of the generator e_2 to I_1 and I_4*

2.4.5. *Contribution of the noise generator i_2*

By replacing Q1 and Q2 by their equivalent circuit and taking into account only the noise generator i_2, we obtain the circuit of Figure 2.6. In this arrangement, the control voltage of the controlled current generator Y_{21} is zero and no current flows into Y_{11}. The control voltage of the controlled current generator Y_{12} is $i_2/(Y_{22}+Y_{33})$. The control voltage of the controlled current generator Y_{43} is also $i_2/(Y_{22}+Y_{33})$. We can therefore write the equations [2.17] and [2.18]:

$$I_1 = \frac{Y_{12}}{Y_{22}+Y_{33}} \cdot i_2 \quad [2.17]$$

$$I_4 = \frac{Y_{43}}{Y_{22}+Y_{33}} \cdot i_2 \quad [2.18]$$

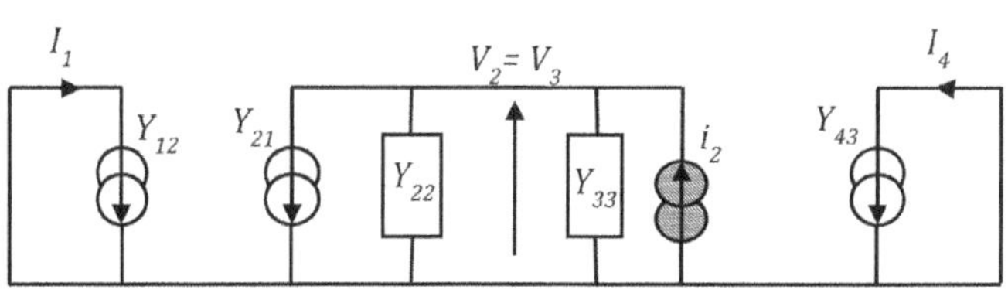

Figure 2.6. *Equivalent diagram of Q1 and Q2 in cascade: contribution of the generator i_2 to I_1 and I_4*

2.5. e_{Tot} and i_{Tot} identification

Using equations [2.5] and [2.11] to [2.18], we can write equations [2.19] and [2.20] that give expressions of I_1 and I_4 for noise to identify e_{Tot} and i_{Tot}:

$$I_1 = Y_{11Tot} \cdot e_1 - i_1 - \left(\frac{Y_{33}Y_{12}}{Y_{22}+Y_{33}}\right) \cdot e_2 + \frac{Y_{12}}{Y_{22}+Y_{33}} \cdot i_2 = Y_{11Tot} \cdot e_{Tot} - i_{Tot} \quad [2.19]$$

$$I_4 = Y_{21Tot} \cdot e_1 + \frac{Y_{22}Y_{43}}{Y_{22}+Y_{33}} \cdot e_2 + \frac{Y_{43}}{Y_{22}+Y_{33}} \cdot i_2 = Y_{21Tot} \cdot e_{Tot} \quad [2.20]$$

By using equation [2.8] in equation [2.20], we obtain the expression of the voltage e_{Tot} given by [2.21]:

$$e_{Tot} = e_1 - \frac{Y_{22}}{Y_{21}} \cdot e_2 - \frac{1}{Y_{21}} \cdot i_2 \quad [2.21]$$

By using equations [2.21] and [2.7] in equation [2.19], we obtain the expression of the current i_{Tot} given by [2.22]:

$$i_{Tot} = i_1 - \frac{(\Delta_y \cdot e_2 + Y_{11} \cdot i_2)}{Y_{21}} \quad [2.22]$$

Where: $\Delta_y = Y_{11}Y_{22} - Y_{21}Y_{12}$

2.6. Calculation of $F_{12}(YS)$

Using equation [2.3] with equations [2.21] and [2.22], we can express $F_{12}(Y_S)$ the cascading Q1, Q2 association noise factor from the noise generators e_1, i_1, e_2 and i_2 for a source admittance Y_S. The expression of $F_{12}(Y_S)$ is given by equation [2.23]:

$$F_{12}(Y_S)_{lin} = 1 + \frac{\left| Y_S . \left(e_1 - \frac{1}{Y_{21}} . (Y_{22} . e_2 + i_2) \right) + i_1 - \frac{\Delta_y e_2 + Y_{11} i_2}{Y_{21}} \right|^2}{|\bar{i_S}|^2} \quad [2.23]$$

Since the noise generators of Q1 and those of Q2 are not correlated, equation [2.23] can be rewritten in the form given by equation [2.24]:

$$F_{12}(Y_S)_{lin} = 1 + \frac{|\overline{Y_S . e_1 + i_1}|^2}{|\bar{i_S}|^2} + \frac{1}{|Y_{21}|^2} \cdot \frac{|\overline{Y_S . (Y_{22} . e_2 + i_2) + \Delta_y e_2 + Y_{11} i_2}|^2}{|\bar{i_S}|^2} \quad [2.24]$$

The noise factor of Q1 is $F_1(Y_S)_{lin} = 1 + \frac{|\overline{Y_S.e_1+i_1}|^2}{|\overline{\iota_S}|^2}$ because Q1 is charged at the input by Y_S. After rearranging the terms in e_2 and i_2, the noise factor $F_{12}(Y_S)$ is written as indicated by equation [2.25] with Y_{OUT1}, the output admittance of the two-port Q1 when its input is charge by Y_S given by equation [2.26]:

$$F_{12}(Y_S)_{lin} = F_1(Y_S)_{lin} + \frac{|Y_S+Y_{11}|^2}{|Y_{21}|^2}.\frac{|\overline{Y_{OUT1}.e_2+i_2}|^2}{|\overline{\iota_S}|^2} \qquad [2.25]$$

$$Y_{OUT1}(Y_S) = Y_{22} - \frac{Y_{12}Y_{21}}{Y_{11}+Y_S} \qquad [2.26]$$

The noise factor of Q2 is $F_2(Y_{OUT1})_{lin} = 1 + \frac{|\overline{Y_{OUT1}.e_2+i_2}|^2}{|\overline{\iota_{OUT1}}|^2}$ because Q2 is terminated at its input by Y_{OUT1} with $|\overline{\iota_{OUT1}}|^2 = 4kTdf \cdot G_{OUT1}$ and where G_{OUT1} is the real part of admittance Y_{OUT1}.

We can therefore rewrite equation [2.25] in the form given by equation [2.27]:

$$F_{12}(Y_S)_{lin} = F_1(Y_S)_{lin} + \frac{|Y_S+Y_{11}|^2}{|Y_{21}|^2}.\frac{G_{OUT1}}{G_S}.(F_2(Y_{OUT1}) - 1) \qquad [2.27]$$

To establish the Friis formula, we only need to identify the term $\frac{|Y_S+Y_{11}|^2}{|Y_{21}|^2}.\frac{G_{OUT1}}{G_S}$ which must be the inverse of the available power gain of the two-port Q1 for a input admittance Y_S and which we see that it concerns only the two-port Q1 and the admittance Y_S which charges its input.

2.7. Friis Formula

2.7.1. *Introduction*

We will therefore calculate the available power gain of the two-port Q1 for a source admittance Y_S. For this, we will recall the definitions of the power transduction gain of a two-port $G_T(Y_S, Y_L)$ and the available power gain of a two-port $G_{Avai}(Y_S)$ and then give the expressions using the flow graph and Mason's rule (Mason 1956).

2.7.2. Transducer power gain

The definition of the transducer power gain of a two-port $G_T(Y_S, Y_L)$ is given by equation [2.28]:

$$G_T(Y_S, Y_L) = \frac{P_L}{P_{SAvai}} \qquad [2.28]$$

Where P_L is the power collected in the charge Y_L and P_{SAvai} is the power available at the source.

To express this gain, we will use two flow graphs: one is given in figure 2.7 to calculate P_{SAvai} and another is shown in Figure 2.8 to calculate P_L.

Figure 2.7 shows the flow graph which describes the connection between a generator consisting of an ideal voltage source E in series with an admittance Y_S and a given charge Y_L.

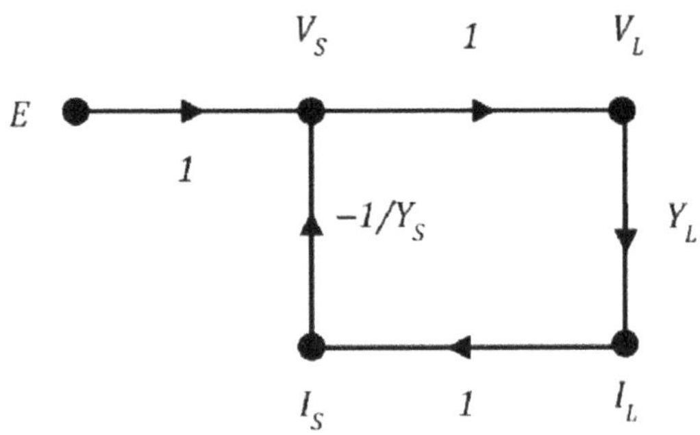

Figure 2.7. *Flow graph for the connection between a source E, Y_S and a charge Y_L*

In fact, this flow graph only describes graphically the relations that exist in this circuit; i.e.:

– $V_L = V_S$: the voltage at the terminals is equal to the output voltage of the generator;

– $I_S = I_L$: the current supplied by the source is equal to the current that flows in the charge;

– $I_L = Y_L \cdot V_L$: Ohm's law regarding the charge;

– $V_S=E-I_S/Y_S$: the output voltage of the generator is equal to the ideal voltage source minus the voltage drop in the internal impedance of the generator.

The power collected in the charge Y_L is $P_L = Re(V_L.I_L^*) = |V_L|^2 \cdot Re(Y_L)$ where $Re(C)$ is the real part operator for the complex C. The voltage V_L in function of E is given by equation [2.29]:

$$V_L = \frac{E}{1+\frac{Y_L}{Y_S}} \quad [2.29]$$

This power P_L will be maximum if $Y_L = Y_S^*$ and the maximum power collected in Y_L is then called P_{SAvai} the power available from the generator. Its expression is given by equation [2.30]:

$$P_{SAvai} = \frac{|E|^2 \cdot |Y_S|^2}{4G_S} \quad [2.30]$$

Where G_S is the real part of the Y_S source admittance.

In Figure 2.8, the flow graph is shown which describes the connection between a source E, Y_S and a two-port which is known by its admittance matrix and which is terminated by any given load Y_L.

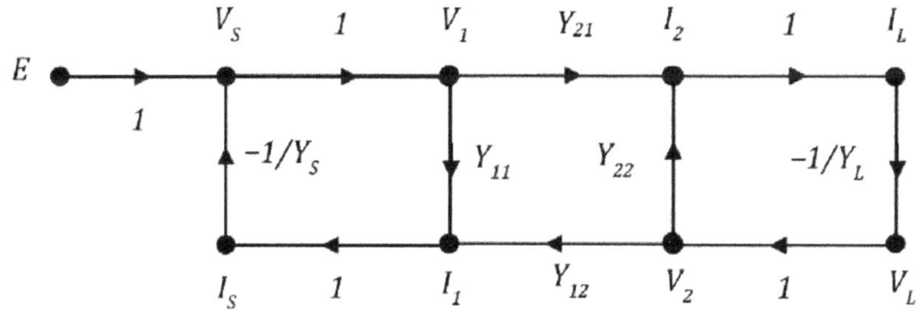

Figure 2.8. *Flow graph of the connection between a source E, Y_S, a two-port Q known by its admittance matrix [Y] and a load Y_L*

Under these conditions, the power collected in the load Y_L is $P_L = |V_L|^2 \cdot Re(Y_L)$ and V_L must be calculated with respect to E using, for example, Mason's rule.

After having identified the three loops – two of which are disjointed – in the graph of Figure 2.8 and applying Mason's rule, we obtain the expression of V_L/E given by the equation [2.31]:

$$\frac{V_L}{E} = \frac{-\frac{Y_{21}}{Y_L}}{1+\frac{Y_{11}}{Y_S}+\frac{Y_{22}}{Y_L}-\frac{Y_{12}Y_{21}}{Y_S\,Y_L}+\frac{Y_{11}Y_{22}}{Y_S\,Y_L}} \quad [2.31]$$

The power collected in the load is given by equation [2.32]:

$$P_L = |E|^2 \cdot \left|\frac{\frac{Y_{21}}{Y_L}}{1+\frac{Y_{11}}{Y_S}+\frac{Y_{22}}{Y_L}-\frac{Y_{12}Y_{21}}{Y_S\,Y_L}+\frac{Y_{11}Y_{22}}{Y_S\,Y_L}}\right|^2 \cdot Re(Y_L) \quad [2.32]$$

From there, we can express the transducer power gain of this two-port using equations [2.28], [2.30] and [2.32]. Its expression is given by equation [2.33a]:

$$G_T(Y_S, Y_L) = \frac{4 \cdot G_L \cdot G_S \cdot |Y_{21}|^2}{|Y_S \cdot Y_L + Y_{11} \cdot Y_L + Y_{22} \cdot Y_S - Y_{21} \cdot Y_{12} + Y_{11} \cdot Y_{22}|^2} \quad [2.33a]$$

Where G_L is the real part of the admittance Y_L.

2.7.3. *Available power gain*

To express the available power gain of a two-port, we must get, in [2.33a], Y_{OUT1} whose expression given by equation [2.26] then let $Y_L=Y_{OUT1}^*$. For the first step, the simplest action is to start from equation [2.31] and rewrite it in the form given by [2.33b]:

$$\frac{I_L}{E} = \frac{Y_{21}Y_S}{Y_S+Y_{11}} \cdot \frac{Y_L}{Y_L+Y_{OUT1}} \quad [2.33b]$$

Under these conditions, the power collected in the charge Y_L is expressed as indicated in the equation [2.34]:

$$P_L = |E|^2 \cdot \left|\frac{Y_{21}Y_S}{Y_S+Y_{11}}\right|^2 \cdot \frac{G_L}{|Y_L+Y_{OUT1}|^2} \quad [2.34]$$

The transducer power gain is written as indicated in equation [2.35]:

$$G_T(Y_S, Y_L) = \left|\frac{Y_{21}}{Y_S+Y_{11}}\right|^2 \cdot \frac{4.G_L \cdot G_S}{|Y_L+Y_{OUT1}|^2} \quad [2.35]$$

To obtain the gain available in power, we need to write G_T for $Y_L=Y_{OUT1}^*$. The result is given by equation [2.36]:

$$G_{Avai}(Y_S) = \left|\frac{Y_{21}}{Y_S+Y_{11}}\right|^2 \cdot \frac{4 \cdot G_{OUT1} \cdot G_S}{|Y_{OUT1}{}^* + Y_{OUT1}|^2} = \left|\frac{Y_{21}}{Y_S+Y_{11}}\right|^2 \cdot \frac{G_S}{G_{OUT1}} \qquad [2.36]$$

We see that this quantity $G_{Avai}(Y_S)$ is that which appears in equation [2.27]. Equation [2.27] can be, finally, rewritten in the form given by equation [2.37] to express the noise factor of two two-ports Q1 and Q2 in cascade:

$$F_{12}(Y_S) = F_1(Y_S) + \frac{F_2(Y_{OUT1})-1}{G_{Avai1}(Y_S)} \qquad [2.37]$$

This equation [2.37] is the Friis formula as stated in the introduction.

2.8. Conclusion

In this chapter, we have demonstrated in a rigorous and explicit way the Friis formula by explaining the quantities involved and under what conditions they must be calculated. The method used, although long due to the many manipulated equations, makes it possible to arrive at the result without conceptual difficulties. It was also an occasion to review the flow graph, Mason's rule and the different definitions that one can give to power gain. An alternative method that could have been used is based on noise power waves.

A good understanding of the Friis formula is essential for the measurement of noise factor because it is this which makes it possible to extract the contribution of the measurement chain, which is also noisy like any electronic device, and thus to go back to the only contribution of the object under test, as we shall see in Chapter 4.

3

Adapted Attenuator and Noise Factor

3.1. Introduction

The objective of this chapter is to show explicitly that the noise figure on 50 Ω, denoted as *NF(50 Ω)* expressed in dB, of a resistive attenuator adapted on 50 Ω, is equal to its attenuation expressed in dB and denoted as $L_{dB} = -10log_{10}|S_{21}|^2$. This translates into a linear equality given by equation [3.1]:

$$F(50\,\Omega)_{lin} = 1/|S_{21}|^2 \tag{3.1}$$

This equation is very interesting because it makes it easy to have a noise factor standard capable of validating a measurement procedure and/or a measurement bench. Indeed, it suffices to have adapted attenuator and measure its gain with a network analyzer.

The calculation hypothesis is that the attenuator and the source which feeds it are at the same temperature $T = T_0 = 290$ K.

To demonstrate this, we shall calculate successively:

– the attenuator's Y and S parameters and the condition for the attenuator to be adapted to 50 Ω;

– equivalent noise generators at the attenuator's input;

– the noise factor on 50 Ω of this attenuator, expressed in linear and denoted as $F(50\,\Omega)_{lin}$.

Then, we will simplify its expression taking into account the adaptation.

3.2. Calculation of Y and S parameters

The resistive attenuator we are going to study is a Π attenuator, composed of three conductances of which two are equal, as shown in Figure 3.1.

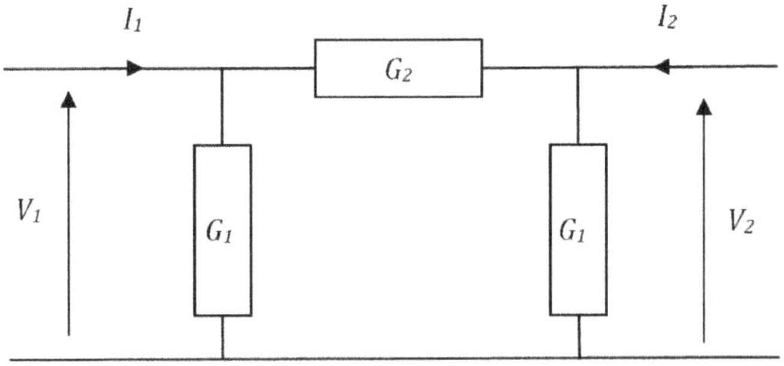

Figure 3.1. *Resistive Π attenuator*

The admittance matrix Y of this attenuator is:

$$\begin{vmatrix} I_1 \\ I_2 \end{vmatrix} = \begin{bmatrix} Y_{11} & Y_{12} \\ Y_{21} & Y_{22} \end{bmatrix} \cdot \begin{vmatrix} V_1 \\ V_2 \end{vmatrix}$$

with equations [3.2] and [3.3]:

$$Y_{11} = Y_{22} = G_1 + G_2 \qquad [3.2]$$

$$Y_{21} = Y_{12} = -G_2 \qquad [3.3]$$

These two terms are obtained by calculating $Y_{11} = \frac{I_1}{V_1}|_{V_2=0}$ *and* $Y_{21} = \frac{I_2}{V_1}|_{V_2=0}$ and using the symmetry of the device.

In order to find the equation linking G_1 and G_2 to get the adaptation on 50 Ω, we just have to write that the input admittance Y_{In} when the output is loaded by the reference conductance $G_0 = 20$ mS is equal to G_0, i.e.:

$$Y_{In} = \frac{I_1}{V_1}|_{G_0} = G_0$$

This equation is written as:

$$Y_{In} = \frac{I_1}{V_1}|_{G_0} = G_1 + \frac{1}{\frac{1}{G_2} + \frac{1}{G_1 + G_0}}$$

hence:

$$G_2 = \frac{G_0^2 - G_1^2}{2G_1} = \frac{(G_0+G_1)\cdot(G_0-G_1)}{2G_1} \qquad [3.4]$$

If equation [3.4] is satisfied, then $S_{11} = S_{22} = 0$. The term S_{21}, equal to S_{12}, is expressed by calculating the composite voltage gain of the circuit of Figure 3.1 when it is connected as input to a source of internal admittance G_0 and of voltage E and at the output to a load of value G_0. The value obtained is given by equation [3.5]:

$$S_{21} = \frac{2\cdot V_2}{E}|_{G_0,G_0} = 2\cdot\frac{1}{2}\cdot\frac{\frac{1}{G_1+G_0}}{\frac{1}{G_2}+\frac{1}{G_1+G_0}} = \frac{G_0 - G_1}{G_0 + G_1} \qquad [3.5]$$

3.3. General representation of noise in two-ports

To represent noise in a noisy two-port, two partially correlated equivalent noise sources can be used (Rothe *et al.* 1956): one in voltage and the other in current, placed at the input of the two-port as shown in Figure 3.2.

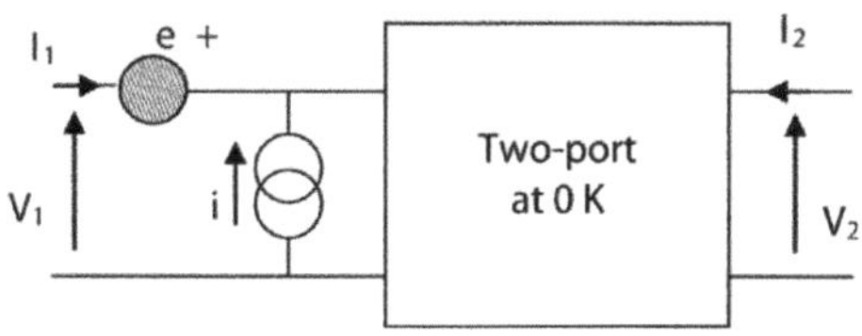

Figure 3.2. *Two-port at 0K with voltage sources and equivalent noise current*

Under these conditions (Friis 1944), the noise factor of this two-port $F(Y_S)$, when a source with an internal admittance Y_S feeds it, is given by equation [3.6] taking into account the arbitrarily chosen directions for *e* and *i*:

$$F(Y_S)_{lin} = 1 + \frac{\overline{|e\cdot Y_S + i|^2}}{\overline{|i_S|^2}} \qquad [3.6]$$

With i_S being the source of noise current of the source admittance.

hence:

$$|\overline{i_S}|^2 = 4kTdf \cdot G_S$$

Equation [3.6] can be established by calculating the two noise powers $N|_{e=i=0}$ (due to i_S only) and $N|_{iS=0}$ (due to e and i) collected in the input admittance of the two-port and by writing $F = 1 + N|_{iS=0} / N|_{e=i=0}$ because i_S is not correlated with e nor with i. The form of equation [3.6] is the main interest of the noise representation in Figure 3.2 where all the noise due to the two-port is brought back on its input.

With this same representation of Figure 3.2, we can also write the matrix form given by the equation [3.7]:

$$\begin{vmatrix} I_1 \\ I_2 \end{vmatrix} = [Y] \cdot \begin{vmatrix} V_1 \\ V_2 \end{vmatrix} + \begin{bmatrix} Y_{11} & -1 \\ Y_{21} & 0 \end{bmatrix} \cdot \begin{vmatrix} e \\ i \end{vmatrix} \qquad [3.7]$$

3.4. Equivalent noise generators at the input of the adapted attenuator

For the Π attenuator, the most natural noise representation is that of Figure 3.3. In this circuit, none of the three noise sources are correlated.

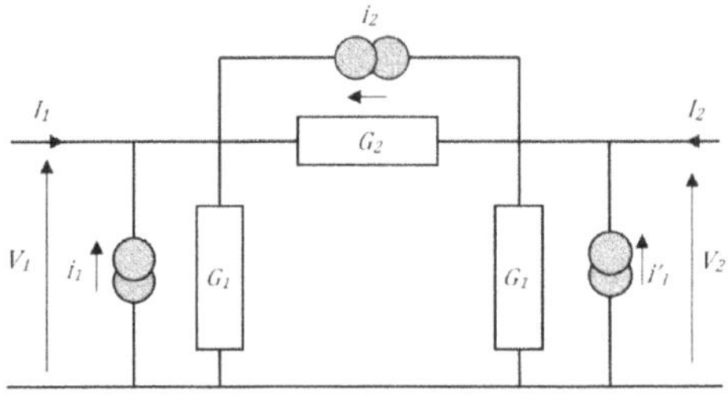

Figure 3.3. *Resistive attenuator in π with noise current sources for each conductance*

To calculate, in accordance with the form of Figure 3.2, the noise generators, one in voltage and the other in current, of the adapted attenuator, we will calculate the contributions of i_1, i'_1 and i_2 to I_1 and I_2 when $V_1 = V_2 = 0$

in the circuit of Figure 3.3 and identify them with the form given by equation [3.7]. It should be noted that the signs attributed to voltage or current noise generators are arbitrary, but as Kirchhoff's laws will be used to establish these equivalences, we must make choices.

The result is as follows:

$I_1 = -i_1 - i_2$

$I_2 = -i'_1 + i_2$

Using equations [3.2] and [3.3], the identification with equation [3.7] gives the expressions of e and i in equations [3.8] and [3.9]:

$$e = -\frac{i_2 - i'_1}{G_2} \quad [3.8]$$

$$i = i_1 - i_2\frac{G_1}{G_2} + i'_1.\frac{G_1 + G_2}{G_2} \quad [3.9]$$

We see in these equations that the equivalent noise generators e and i are partially correlated since i'_1 and i_2 appear in e and i. It is the topology of the circuit that brings this correlation.

3.5. Noise factor on 50 Ω of the adapted attenuator

This calculation is done for $Y_S = G_0$.

Under these conditions, equation [3.6] is written as indicated in equation [3.10]:

$$F(50\,\Omega)_{lin} = 1 + \frac{G_0^2|\bar{e}|^2 + |\bar{\iota}|^2 + 2G_0 \cdot Re(\overline{e\iota^*})}{|\overline{\iota_S}|^2} \quad [3.10]$$

with:

$$|\overline{\iota_S}|^2 = 4kTdf \cdot G_0$$

$$|\bar{e}|^2 = \frac{\overline{|\iota_2 - \iota'_1|^2}}{G_2^2} = 4kTdf.\left[{}^{1}/_{G_2} + {}^{G_1}/_{G_2^2}\right]$$

$$|\overline{\iota}|^2 = 4kTdf \cdot \left[G_1 + {G_1^2}\Big/{G_2} + {(G_1+G_2)^2}\Big/{G_2^2} \right]$$

$$\overline{e \cdot \iota^*} = 4kTdf \cdot \left[{G_1^2}\Big/{G_2^2} + {2G_1}\Big/{G_2} \right]$$

Under these conditions, the attenuator's noise factor is:

$$F(50\,\Omega)_{lin} = 1 + G_0 \cdot \left[\frac{1}{G_2} + \frac{G_1}{G_2^2}\right] + \frac{G_1}{G_0} \cdot \left[2 + \frac{3G_1}{G_2} + \frac{G_1^2}{G_2^2}\right] + 2 \cdot \left[\frac{2G_1}{G_2} + \frac{G_1^2}{G_2^2}\right]$$

$$F(50\,\Omega)_{lin} = 1 + 2 \cdot \frac{G_1}{G_0} + \frac{1}{G_2} \cdot \left[G_0 + \frac{3G_1^2}{G_0} + 4G_1\right] + \frac{1}{G_2^2} \cdot \left[G_0G_1 + \frac{G_1^3}{G_0} + 2 \cdot G_1^2\right]$$

$$F(50\,\Omega)_{lin} = 1 + 2 \cdot \frac{G_1}{G_0} + \frac{1}{G_0G_2} \cdot [G_0^2 + 3G_1^2 + 4G_0G_1] + \frac{G_1}{G_0G_2^2} \cdot [G_0 + G_1]^2$$

$$F(50\,\Omega)_{lin} = 1 + 2 \cdot \frac{G_1}{G_0} + \frac{1}{G_0G_2} \cdot [(G_0 + 2G_1)^2 - G_1^2] + \frac{G_1}{G_0G_2^2} \cdot [G_0 + G_1]^2$$

$$F(50\,\Omega)_{lin} = 1 + 2 \cdot \frac{G_1}{G_0} + \frac{(G_0 + G_1) \cdot (G_0 + 3G_1)}{G_0G_2} + \frac{G_1}{G_0G_2^2} \cdot [G_0 + G_1]^2$$

At this stage of the calculation, we can replace G_2 by its expression as a function of G_0 and G_1 by noticing that:

$$G_2^2 = \frac{(G_0 + G_1)^2 \cdot (G_0 - G_1)^2}{4G_1^2}$$

$$F(50\,\Omega)_{lin} = 1 + 2 \cdot \frac{G_1}{G_0} + \frac{2G_1}{G_0} \cdot \left[\frac{G_0 + 3G_1}{G_0 - G_1}\right] + \frac{4G_1^3}{G_0[G_0 - G_1]^2}$$

$$F(50\,\Omega)_{lin} = 1 + 2 \cdot \frac{G_1}{G_0}\left[1 + \frac{G_0 + 3G_1}{G_0 - G_1} + \frac{2G_1^2}{[G_0 - G_1]^2}\right]$$

$$F(50\ \Omega)_{lin} = 1 + 2 \cdot \frac{G_1}{G_0 \cdot [G_0 - G_1]^2}[[G_0 - G_1]^2 + (G_0 + 3G_1) \cdot (G_0 - G_1) + 2G_1^2]$$

The final result is given by equation [3.11]:

$$F(50\ \Omega)_{lin} = 1 + 2 \cdot \frac{2.G_1 G_0^2}{G_0 \cdot [G_0 - G_1]^2} = 1 + \frac{4G_1 G_0}{[G_0 - G_1]^2} = \frac{[G_0 + G_1]^2}{[G_0 - G_1]^2} = {}^{1}/_{|S_{21}|^2} \qquad [3.11]$$

3.6. Using Bosma's theorem

Another way to demonstrate this result is to use noise power waves and Bosma's theorem.

Under these conditions and as shown in Figure 3.4, the noisy two-port (Heicken 1981) is described by:

– its matrix S and the power waves incident and reflected on its ports 1 and 2;

– a noise power wave b_{n1} that emerges from port 1;

– a noise power wave b_{n1} that emerges from port 2.

The input load is described by its reflection coefficient Γ_S and a noise power wave b_{nS}.

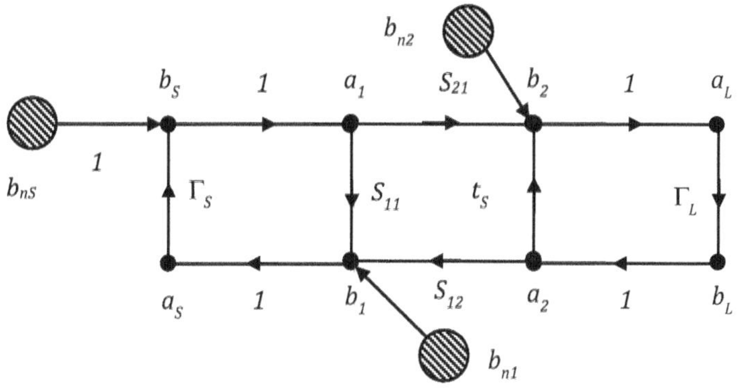

Figure 3.4. *Noise and power waves*

The output load assumed to be non-noisy as always for the calculation of the noise factor is described by its reflection coefficient Γ_L.

Using the flow graph in Figure 3.4, we can calculate the noise factor of the two-port by comparing the noise power collected in Γ_L taking into account b_{nS}, b_{n1} and b_{n2} and that collected in Γ_L by holding only count of b_{nS}. Since b_{nS} is not correlated with b_{n1} nor with b_{n2}, we can write the noise factor in the form given by equation [3.12]:

$$F(\Gamma_S)_{lin} = 1 + \frac{(|\overline{a_L}|^2 - |\overline{b_L}|^2)_{b_{nS}=0}}{(|\overline{a_L}|^2 - |\overline{b_L}|^2)_{b_{n1}=b_{n2}=0}} \quad [3.12]$$

In this case, we calculate for $\Gamma_S = 0$ because we look for the noise factor for a source impedance equal to $R_0 = 50\Omega$. Strictly speaking, the powers collected in the charge must be calculated as available powers; that is, for a reflection coefficient charge Γ_L equal to the conjugate of the reflection coefficient seen at the output of the attenuator, when its input is charged by Γ_S. As the attenuator is adapted on its input and on its output ($S_{11} = S_{22} = 0$) and $\Gamma_S = 0$, it is just necessary to make $\Gamma_L = 0$.

Under these conditions, the graph of Figure 3.4 is reduced to that of Figure 3.5.

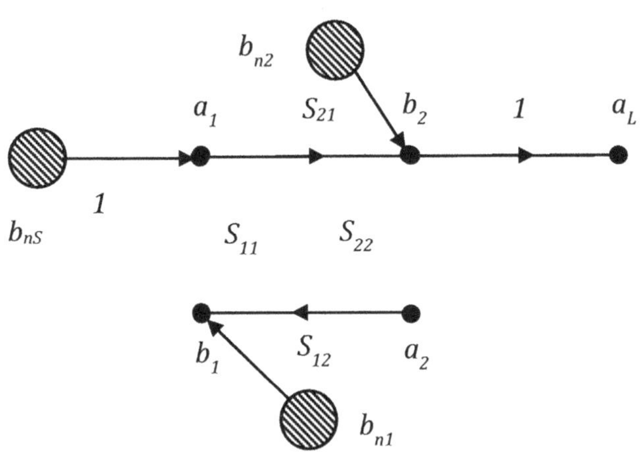

Figure 3.5. *Flow graph for $\Gamma_S = \Gamma_L = 0$*

Equation [3.12] becomes that given by the equation [3.13]:

$$F(\Gamma_S = 0)_{lin} = 1 + \frac{(|\overline{a_L}|^2)_{b_{nS}=0}}{(|\overline{a_L}|^2)_{b_{n1}=b_{n2}=0}} \qquad [3.13]$$

Using the flow graph in Figure 3.5, we obtain equations [3.14] and [3.15]:

$$(|\overline{a_L}|^2)_{b_{nS}=0} = |\overline{b_{n2}}|^2 \qquad [3.14]$$

$$(|\overline{a_L}|^2)_{b_{n1}=b_{n2}=0} = |S_{21}|^2 \,.\, |\overline{b_{nS}}|^2 \qquad [3.15]$$

We use Bosma's theorem to calculate these two terms.

Bosma's theorem (Bosma 1967) shows that at the thermodynamic equilibrium for a passive device with i ports, the correlation matrix of the noise power waves is written as:

$$[B_n \cdot B_n^\tau] = kTdf \cdot [I - S \cdot S^\tau] \qquad [3.16]$$

where:

– S is the matrix i-rows i-columns S of the object and S^τ is its conjugated transpose;

– I is the identity matrix of rank i;

– B_n is the i-rows 1-column matrix of noise power waves and conjugated transpose $B_n{}^\tau$. The product is a matrix i-rows i-columns.

We obtain

$$(|\overline{a_L}|^2)_{b_{nS}=0} = |\overline{b_{n2}}|^2 = kTdf \cdot (1 - |S_{21}|^2) \qquad [3.17]$$

$$(|\overline{a_L}|^2)_{b_{n1}=b_{n2}=0} = |\overline{b_{nS}}|^2 \cdot |S_{21}|^2 = kTdf \cdot |S_{21}|^2 \qquad [3.18]$$

By using equations [3.17] and [3.18] respectively in [3.14] and [3.15] and in equation [3.13], we obtain the expression given by equation [3.19] which corresponds well to the sought form:

$$F(\Gamma_S = 0)_{lin} = 1/|S_{21}|^2 \qquad [3.19]$$

3.7. Consequences on the structure of a receiver

Let us consider a reception chain comprising an amplifier adapted on its two accesses, of available gain $G_{Avai1}(\Gamma_S = 0)_{dB} = 17\ dB$ and of noise factor $NF_1(\Gamma_S = 0)_{dB} = 6\ dB$, and of an adapted attenuator of available gain $G_{Avai2}(\Gamma_S = 0)_{dB} = -10\ dB$ and therefore of noise factor $NF_2(\Gamma_S = 0)_{dB} = 10\ dB$.

Using the Friis formula given by equation [2.37], we can study the noise factor according to the way in which these two stages are cascaded.

If we do the cascade combination attenuator at the head then amplifier, the noise factor on 50 Ω will be:

$$F_{21}(\Gamma_S = 0)_{lin} = 10 + \frac{(4-1)}{0.1} = 40 \text{ or } NF_{21}(\Gamma_S = 0) = 16 \text{ dB}$$

If the cascade is combined amplifier at the head then the attenuator, the noise factor on 50 Ω will be:

$$F_{12}(\Gamma_S = 0)_{lin} = 4 + \frac{(10-1)}{50} \#4.2 \text{ let } NF_{12}(\Gamma_S = 0) = 6.2 \text{ dB}$$

So we see that in the second case, the noise factor of the attenuator is almost completely hidden by the amplifier at the head of the receiver while in the first case, the noise factor is much greater.

As a first consequence, we can say that when designing a reception chain, we will put, as close as possible to the input, a low-noise and high-gain amplifier that will act as a “mask”, as much as possible, towards the noise factor of the following stages. It will however be necessary to ensure that this 1st stage does not saturate at its output; which would compromise the further processing of the useful signal received.

Moreover, in a reception chain, it is often necessary to put a pre-selection filter at the input, which is to say before the low noise amplifier, to eliminate unwanted signals, in particular the image frequency. It will therefore be necessary to ensure that the losses of this filter are as low as possible in the receiver’s bandwidth.

The second consequence explains, in part, the high noise factor of spectrum analyzers. Indeed, in these devices, the input drives an adjustable attenuator for protecting the 1st stage of the spectrum analyzer which is a mixer. It will therefore be necessary, when using a spectral analyzer to make noise measurements:

– turn off the input attenuator but being careful not to apply high power signals to the analyzer input;

– activate the internal preamplifier of the spectrum analyzer if one exists;

– or put, closer to the object to be measured, a low noise amplifier.

If a cable is needed to make the connection, it is better to put it between the output of the low noise amplifier and the spectrum analyzer than between the device under test and the input of the low noise amplifier; this to avoid degrading the noise factor of the measuring chain due to cable losses.

3.8. Equivalent noise resistance and noise conductance at the input

Expressing these two quantities was not essential for the demonstration, but it is simply a matter of highlighting, in this example, what are the equivalent noise resistance and the equivalent noise conductance at the input of the attenuator and the lack of a relationship between these two quantities, as announced in section 1.4.3 of Chapter 1.

Taking again the expression [3.8] for e and the definition [1.12b] of R_N, we obtain, for our attenuator, the expression of R_N, given by equation [3.20]:

$$R_N = \frac{G_1+G_2}{G_2^2} \qquad [3.20]$$

Using [1.12c], we can show that the correlation admittance is given by [3.21]:

$$Y_{cor} = \frac{\overline{\iota \cdot e^*}}{\overline{|e|^2}} = \frac{(G_1+2 \cdot G_2) \cdot G_1}{G_1+G_2} \qquad [3.21]$$

Using again [1.12c], we can express i_n and calculate, from its definition [1.12c], the value of G_n. It is given by [3.22]:

$$\mathrm{G_n} = G_1 + \frac{G_1 \cdot G_2}{G_1+G_2} = \frac{(G_1+2 \cdot G_2) \cdot G_1}{G_1+G_2} \qquad [3.22]$$

For R_N and G_n to be the opposite of each other, the equality [3.23] must be verified:

$$G_2{}^2 = G_1{}^2 + 2 \cdot G_1 \cdot G_2 \qquad [3.23]$$

However, for the adaptation of the attenuator, we must have equation [3.4] linking the three conductances of the circuit: $2 \cdot G_1 \cdot G_2 = G_0{}^2 - G_1{}^2$. So we need $G_2{}^2 = G_0{}^2$.

If we choose the positive root for G_2, which seems reasonable since it is a true conductance, then G_1must verify the equation $G_1{}^2 + 2 \cdot G_1 \cdot G_0 - G_0{}^2$.

This equation has a positive root $G_1 = G_0(1 + \sqrt{2})$.

However, G_1 must be less than G_0 according to equation [3.4]. So, it is not possible to have $G_n = 1/R_N$.

3.9. Conclusion

The noise factor over 50 Ω of an adapted attenuator is indeed $F(50\ \Omega)_{lin} = 1/|S_{21}|^2$.

This result could have been found without making any calculation by making the following remarks:

– the attenuator being input adapted and symmetrical, $S_{22} = 0$ is an output admittance equal to G_0 when the input is closed on G_0;

– the attenuator, seen from the output, therefore behaves like a conductance G_0 placed at the temperature T;

– the noise power available at the output of this attenuator is $N_{Out} = kTdf$;

– the noise power available at the terminals of the source admittance G_0 placed at the temperature T is $N_{In} = kTdf$.

The noise factor on 50 Ω is therefore:

$$F(50\Omega) = \frac{S_{In}/N_{in}}{S_{Out}/N_{Out}} = \frac{S_{In}}{S_{Out}} \cdot \frac{N_{Out}}{N_{in}} = \frac{S_{In}}{S_{Out}} = 1/|S_{21}|^2$$

In conclusion, it must be remembered that anything that causes losses degrades the signal-to-noise ratio of a signal to be processed and that these losses, often caused by passive devices, must therefore be as low as possible.

4

Noise Factor Measurement on 50 Ω

4.1. Introduction

The purpose of this chapter is to show the procedure for performing noise factor measurements for a source impedance $Z_S = R_0 = 50\ \Omega$. The source impedance value is not a choice, but it is the only possible value if a tuner is unavailable to vary the value of the source impedance. These measurements are done with the following equipment:

– a Keysight N9320A spectrum analyzer with the built-in amplifier option that is used to measure noise power;

– a Keysight 364C noise diode that simulates two source temperatures;

– a ZHL-1010 + Mini-Circuits amplifier to lower the factor of noise of the measurement chain which, even with the amplifier integrated in the spectrum analyzer, remains high;

– the continuous power supplies needed to supply the 28VDC to the noise diode and the 12VDC to the Mini-Circuits amplifier;

– the device under test (DUT), which is to measure the noise factor which is an adapted attenuator 6 dB with SMA connectors.

This measurement bench is shown in Figure 4.1. The advantage of taking an adapted attenuator 6 dB as the device under test is that we know in advance its noise factor on 50 Ω which is 6 dB (Bosma 1967); this makes it possible to check the consistency of the measurements. These measurements will be made by performing a 2nd stage correction consisting of the Mini-

Circuits amplifier and the spectrum analyzer using the Friis formula (Friis 1944).

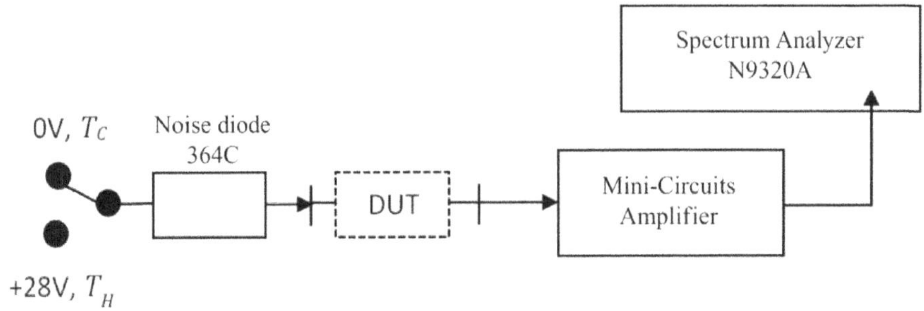

Figure 4.1. *Synopsis of the noise measurement bench*

4.2. Noise factor measurement by Y factor

When making a noise factor measurement for a given source impedance with a noise diode, it is necessary to measure two noise powers N_C and N_H which respectively correspond to the noise power measured at the output of the DUT, when the noise diode is at T_C, and the noise power measured at the output of the DUT, when the noise diode is at T_H.

A noise diode consists mainly of a diode that can be biased at two voltages $V_D = 0$ V or $V_D = -28$ V, its bias circuit and an attenuator connected to the noise output side (Rota 2007).

When the diode is biased at $V_D = 0$ V, the assembly behaves like a load 50 Ω at the temperature $T_C = T_{Amb}$ where T_{Amb} is the physical temperature of the diode. The assembly then delivers an available noise power kT_C in a frequency band of 1 Hz.

When the diode is biased at $V_D = -28$ V, the diode operates in its avalanche zone, generating a significant amount of noise. In this case, the assembly behaves like a load 50 Ω at the temperature T_H, very large compared to T_C, delivering an available noise power kT_H in a frequency band of 1 Hz. Thanks to the attenuator, it is possible to control T_H and thus cause the impedance of the whole remains the same for both temperatures.

The two noise powers N_C and N_H, measured at the spectrum analyzer, are therefore natively expressed in dBm and must therefore be converted to mW. In linear, we can then write equations [4.1] and [4.2] where N_A is the noise power added by the DUT measured at its output, G_{DUT} is the power gain of the DUT, k is the Boltzman constant and *df* the frequency range on which the measurement is made. We will explain later what exactly this G_{DUT} gain is:

$$N_C = kT_C \cdot df \cdot G_{DUT} + N_A \quad [4.1]$$

$$N_H = kT_H \cdot df \cdot G_{DUT} + N_A \quad [4.2]$$

Under these conditions, the noise factor of the DUT, expressed in linear and given by equation [4.3], can be rewritten in the form of equation [4.4], where T_0 is the reference temperature for the noise factor, 290 K:

$$F = 1 + N_A/kT_0 \cdot df \cdot G_{DUT} \quad [4.3]$$

$$F = \frac{Y\left(1 - \frac{T_C}{T_0}\right) + \frac{T_H}{T_0} - 1}{Y - 1} \quad [4.4]$$

with: $Y = N_H/N_C$.

If temperature T_C is equal to T_0, which is true when measurements are made at ambient temperature close to 290 K, then equation [4.4] is written in the form given by equation [4.5]:

$$F = \frac{\frac{T_H - T_0}{T_0}}{Y - 1} \quad [4.5]$$

By using ENR_{lin}, the diode's id *Excess Noise Ratio* which, by definition (Keysight 2017a), is equal to the ratio $\frac{T_H - T_0}{T_0}$, we obtain for the noise factor the definitive expression given by equation [4.6]:

$$F = \frac{ENR_{lin}}{Y - 1} \quad [4.6]$$

In the end, to find F, we must therefore measure both N_H and N_C noise powers and know the ENR_{lin} of the diode used; ENR which is always given by the manufacturer in dB. It is therefore necessary to convert the ENR of the diode to linear using formula [4.7]:

$$ENR_{lin} = 10^{\frac{ENR_{dB}}{10}} \qquad [4.7]$$

4.3. Second stage correction

During the measurement, the DUT and the noise diode are not the only contributors to the noise powers, measured by the spectrum analyzer and if we denote the following:

– N_{A1} as the noise power added by the DUT and measured at its output;

– G_1 as the power gain in of the DUT;

– N_{A2} as the noise power added by the measurement system (Mini-Circuits amplifier and spectrum analyzer) and measured at its output;

– G_2 as the power gain of the measurement chain.

We obtain the expressions given by equations [4.8] and [4.9]:

$$N_C=kT_C{\cdot}df{\cdot}G_1{\cdot}G_2+ N_{A1}{\cdot}G_2 +N_{A2} \qquad [4.8]$$

$$N_H=kT_H{\cdot}df{\cdot}G_1{\cdot}G_2+ N_{A1}{\cdot}G_2 +N_{A2} \qquad [4.9]$$

It is therefore clear that we don't measure $F_1(Z_S)$ the noise factor of the DUT but $F_{12}(Z_S)$ the noise factor of the entire system device under test + measurement chain. This problem was formulated by (Friis 1944) and is described by equation [4.10]:

$$F_{12}(Z_S) = F_1(Z_S) + \frac{F_2(Z_{OUT1})-1}{G_{Avai1}(Z_S)} \qquad [4.10]$$

where:

– F_{12} is the noise factor of the cascaded association of the DUT and the measurement chain and is related to the source impedance seen by the DUT, $F_{12}(Z_S)$;

– F_1 is the noise factor of the DUT and is related to the source impedance seen by this first two-port, $F_1(Z_S)$;

– F_2 is the noise factor of the measurement chain and is relative to the source impedance seen by this two-port Q2, i.e. $F_2(Z_{OUT1})$; Z_{OUT1} being the output impedance of the DUT when its input is loaded by Z_S;

– G_{Avai1} is the available power gain of the DUT and is therefore the transducer power gain calculated for a source impedance Z_S and for a load impedance $Z_L = Z_{OUT1}^*$.

The form given by equation [4.10] is complicated to handle because of $F_2\ (Z_{OUT1})$ and $G_{Avai1}(Z_S)$, but in the case which interests us, it can be simplified because:

– the noise diode has an impedance of 50 Ω for both temperatures;

– the measuring chain is adapted to 50 Ω input;

– the DUT is an attenuator adapted to50 Ω and thus its output impedance Z_{OUT1} is equal to 50 Ω when its input is loaded by 50 Ω.

In these conditions, we have:

– $F_1(Z_S)$=$F_1(R_0)$;

– $Z_{OUT1} = R_0$;

– the gain G_1 brought by the DUT is G_{Avai1}; we will come back later to this problem;

– $F_2(Z_{OUT1}) = F_2(R_0)$, which can be evaluated with the procedure described in section 4.2.

Under these conditions, it is sufficient to measure $F_2(R_0)$ and $F_{12}(R_0)$ and then determine$G_{Avai1} = G_1$ by an additional calculation. This procedure is described in section 4.4.

4.4. Measurement procedure and calculation of the noise factor of the DUT

The procedure takes place in four stages:

– the measurement of $F_2(R_0)$ which we will call calibration of the noise measurement chain and which is done by using the assembly of Figure 4.2 and by measuring the two noise powers N_{CalC} and N_{CalH};

– the measurement of $F_{12}(R_0)$ which is done using the assembly of Figure 4.3 and measuring the two noise powers N_{MesC} and N_{MesH};

– the determination of G_{Avai1} from the previous measurements; calculating $F_1(R_0)$ using equation [4.10].

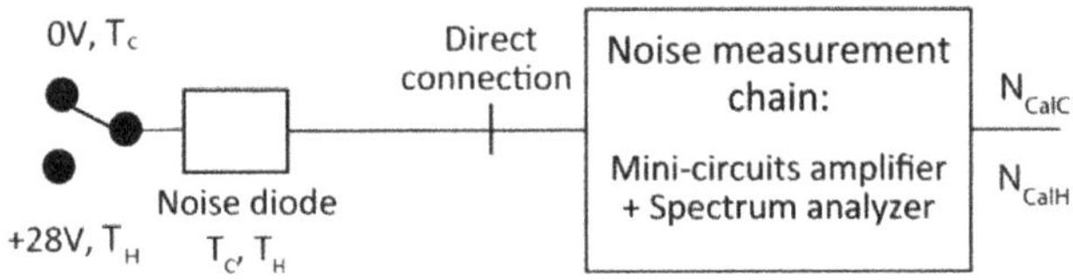

Figure 4.2. *Calibration of the noise measurement chain*

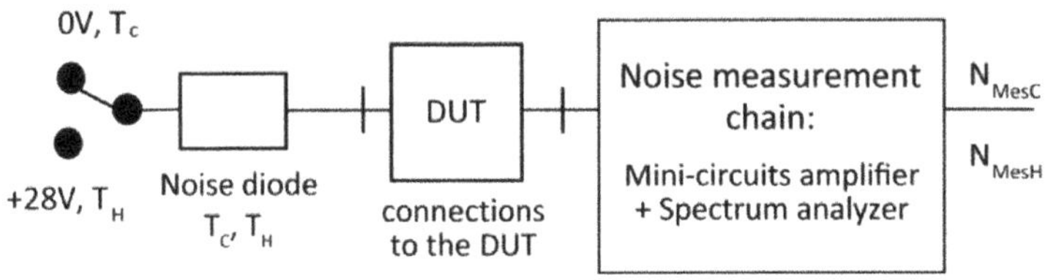

Figure 4.3. *Measurement of noise of the DUT + measurement chain*

Applying equation [4.6] to $Y_{Cal} = N_{CalH}/N_{CalC}$, we obtain the noise factor of the measurement chain:

$$F_2(R_0) = \frac{ENR_{lin}}{Y_{Cal}-1} \quad [4.11]$$

With the same calculation for $Y_{Mes} = N_{MesH}/N_{MesC}$, we obtain the noise factor of the DUT + measurement chain:

$$F_{12}(R_0) = \frac{ENR_{lin}}{Y_{Mes}-1} \quad [4.12]$$

During the calibration phase, using equations [4.1] and [4.2], we also obtain the gain G_2 of the measurement chain by equation [4.13]:

$$G_2 = \frac{N_{CalH}-N_{CalC}}{k(T_H-T_C)\cdot df} \quad [4.13]$$

During the measurement phase, using equations [4.8] and [4.9] and knowing G_2, we obtain the gain G_1 of the DUT by equation [4.14]:

$$G_1 = \frac{N_{MesH} - N_{MesC}}{N_{CalH} - N_{CalC}} \qquad [4.14]$$

As in the measurement conditions, we have the equality $G_{Avai1} = G_1$; [4.10] is then used to calculate $F_1(R_0)$as indicated in equation [4.15]:

$$F_1(R_0) = \frac{ENR_{lin}}{Y_{Mes-1}} - \frac{N_{CalH} - N_{CalC}}{N_{MesH} - N_{MesC}} \cdot \left(\frac{ENR_{lin}}{Y_{Cal-1}} - 1\right) \qquad [4.15]$$

It should be remembered that the use of this expression is valid only if the following four conditions are met:

- $F_1(Z_S) = F_1(R_0)$;
- $Z_{OUT1} = R_0$;
- $G_1 = G_{Avai1}$;
- $F_2(Z_{OUT1}) = F_2(R_0)$.

Of these four conditions, only the condition $G_1 = G_{Avai1}$ has not been clearly demonstrated; we will do this in section 4.5.

4.5. Available power gain and insertion power gain of the DUT

We will give a clear definition of the gain G_1 that appears in equation [4.14]. To evaluate this gain, we use the assembly of Figure 4.2 and then that of Figure 4.3.

The flow graphs of the assemblies of Figures 4.2 and 4.3 are given respectively in Figures 4.4 and 4.5.

Figure 4.4 is the flow graph of the calibration step and the whole measurement chain is represented by its reflection coefficient Γ_L. The noise source (i.e. the noise diode) is represented by the power wave it generates $b_{NH,C}$ and its reflection coefficient Γ_S.

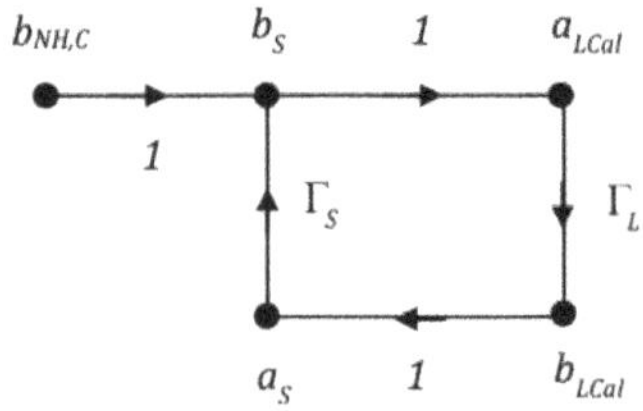

Figure 4.4. *Flow graph during calibration*

Figure 4.5 represents the flow graph of the measurement step. The measurement chain is always represented by its reflection coefficient Γ_L and the source of noise by the power wave that it generates $b_{NH,C}$ and its reflection coefficient being Γ_S. Parameters S in Figure 4.5 are those of the DUT.

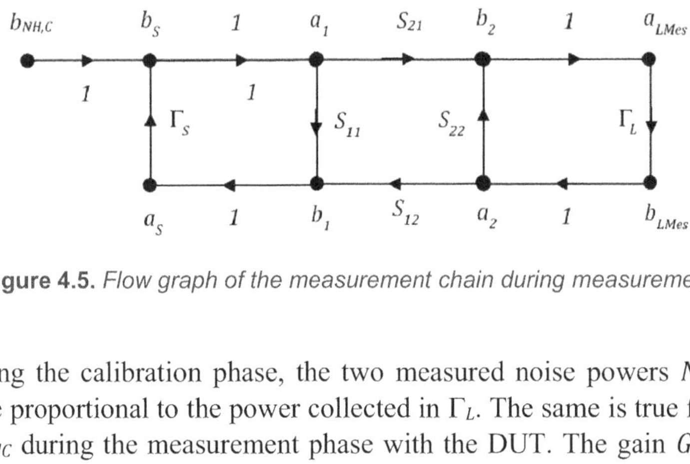

Figure 4.5. *Flow graph of the measurement chain during measurement*

During the calibration phase, the two measured noise powers N_{CalH} and N_{CalC} are proportional to the power collected in Γ_L. The same is true for N_{MesH} and N_{CalC} during the measurement phase with the DUT. The gain G_1 is thus given by the equation [4.16]; it is called the insertion power gain of the DUT because it is the ratio between the power collected in the load when the DUT is inserted between the source and the load and the power collected in the load, when the source is directly connected to the load. Its expression can be established using Mason's rule in the flow graphs of Figures 4.4 and 4.5:

$$G_1 = \frac{|a_{LMes}|^2 \cdot (1-|\Gamma_L|^2)}{|a_{LCal}|^2 \cdot (1-|\Gamma_L|^2)} = \frac{|a_{LMes}|^2}{|a_{LCal}|^2} = \frac{|S_{21}|^2 \cdot |1-\Gamma_L.\Gamma_S|^2}{|(1-S_{11}.\Gamma_S)\cdot(1-S_{22}\cdot\Gamma_L)-S_{12}\cdot\Gamma_S\cdot S_{21}\cdot\Gamma_L|^2} \quad [4.16]$$

At the same time, the available power gain G_{Avai1} is obtained by calculating the transducer power gain of the DUT for a load $\Gamma_L = S_{22}'^*$ where

$$S'_{22} = S_{22} + \frac{S_{21}.S_{12}.\Gamma_S}{1-S_{11}.\Gamma_S}.$$

Recall that the transducer power gain is the ratio of the power collected in the load $P_L = |a_L|^2 \cdot (1 - |\Gamma_L|^2)$ on the power available at the source $P_{SAvai} = |b_{NH,C}|^2/(1 - |\Gamma_S|^2)$.

The expression of the transducer power gain is given by equation [4.17]. It can be established using Mason's rule in the flow graphs of Figures 4.4 and 4.5:

$$G_{T1} = \frac{|S_{21}|^2 \cdot (1-|\Gamma_S|^2) \cdot (1-|\Gamma_L|^2)}{|(1-S_{11} \cdot \Gamma_S) \cdot (1-S_{22} \cdot \Gamma_L) - S_{12} \cdot \Gamma_S \cdot S_{21} \cdot \Gamma_L|^2} \qquad [4.17]$$

By showing S'_{22} in [4.17], we obtain the expression given by equation [4.18] for the transducer power gain:

$$G_{T1} = \frac{|S_{21}|^2 \cdot (1-|\Gamma_S|^2) \cdot (1-|\Gamma_L|^2)}{|(1-S_{11} \cdot \Gamma_S) \cdot (1-S'_{22}.\Gamma_L)|^2} \qquad [4.18]$$

The expression of the available power gain, given by the equation [4.19], is then obtained by making $\Gamma_L = S'^{*}_{22}$ in equation [4.18]:

$$G_{Avai1} = \frac{|S_{21}|^2 \cdot (1-|\Gamma_S|^2)}{|1-S_{11} \cdot \Gamma_S|^2 \cdot \left(1-|S'_{22}|^2\right)} \qquad [4.19]$$

We must therefore compare G_{Avai1} given by equation [4.19] to G_1 given by equation [4.16] in the situation where $\Gamma_L = \Gamma_S = 0$. For the insertion power gain, we obtain the value given by [4.20] and for the available power gain, we obtain that given by equation [4.21]:

$$G_1 = |S_{21}|^2 \qquad [4.20]$$

$$G_{Avai1} = \frac{|S_{21}|^2}{1-|S'_{22}|^2} \qquad [4.21]$$

We see then that it is necessary that S'_{22} be zero; that is true in our case. In all other cases, $F_1(R_0)$ is underestimated by using G_1 in equation [4.10] instead of G_{Avai1} because $G_1 < G_{Avai1}$.

4.6. Sample results

All that remains is to use equation [4.15] on our adapted attenuator; which will allow us to find its noise factor on 50 Ω and its gain.

The settings of the spectrum analyzer are as follows:

– center frequency; that is to say the one where the measurement is made: 1 GHz;

– resolution band: 1 MHz.

If we choose a lower resolution band, we will improve the noise factor of the measurement chain, but also reduce the noise power measured, which will not facilitate accurate measurement. This compromise – a few MHz of resolution band – is the one usually retained (Keysight 2017a):

– frequency span: 10 kHz;

– video filter: 10 kHz;

– input attenuation: 0 dB;

– internal amplifier: ON;

– reference level: – 90 dBm;

– scale of levels: 2 dB / div;

– average power over 50 sweeps.

The Keysight noise diode has an ENR of 14.74 dB at 1 GHz.

The measurement results are given in Table 4.1. These noise power measurements have been rounded to the nearest 1/10 of a dB, even if the display gives a 1/100 of a dB reading.

Diode	Calibration	Measurement
Cold	– 97.2 dBm	– 97.2 dBm
Hot	– 92.4 dBm	– 95.4 dBm

Table 4.1. *Measurement results on a 6 dB attenuator*

These measurement results were processed with an Excel routine to perform the various calculations as shown in Figure 4.6. By observing the green boxes, we obtain:

$$F1 = 5.95 \text{ dB}, G1 = -5.95 \text{ dB}$$

Note that the result is that expected since we have F1(dB) = – G1(dB)#6dB. On the other hand, if we change the measurement results a little, especially if N_{CalC}NCal, C and N_{CalH}NMes, C are different – a few hundredths of dBm of difference are enough – then the results are much less satisfactory.

In any case, these two noise powers must be equal if the DUT, which is passive, is at the same ambient temperature as the noise diode and adapted to its ports (Bosma 1967). A difference between these two powers is the sign that the reflection coefficient of the diode and the reflection coefficient at the output of the attenuator, input loaded by the diode, are different. This discrepancy can easily be seen in the measurements and can be traced back to the origin.

To use this calculation routine, at the measurement frequency, enter the following 5 values:

Noise diode's ENR in dB:		14.74
	The measured noise power without DUT for Tc in dBm:	-97.2
CAL	The measured noise power without DUT for Th in dBm:	-92.4
	The measured noise power with DUT for Tc in dBm:	-97.2
MES	The measured noise power with DUT for Th in dBm:	-95.4

	ENR(dB)	ENR(lin)			
	14.74	29.78516429			
	CAL			MES	
Tc	-97.2	1.90546E-10	Tc	-97.2	1.90546E-10
Th	-92.4	5.7544E-10	Th	-95.4	2.88403E-10
Y		3.01995172			1.513561248
Y-1		2.01995172			0.513561248
	F2(lin)	14.74548327		F12(lin)	57.99729708
	F2(dB)	11.68659011		F12(dB)	17.63407754
	G1(lin)	0.254244318			
	G1(dB)	-5.94748743			
	F1(lin)	3.93322457			
	F1(dB)	5.947487434			

Figure 4.6. *Processing of measurement results by Excel. For a color version of this figure, see www.iste.co.uk/fouquet/electronics.zip*

4.7. Conclusion

With regard to the value of the noise factor of the adapted attenuator, it is possible to compare measurements of parameters S made to the network analyzer. Measured with a full two-port calibrated network analyzer, the 6 dB attenuator gives the following results:

– transmission coefficient: $|S_{21}|^2 = -5.9$ dB;

– reflection coefficient: $Mag(S_{22}) = 0.025$.

The error made by replacing the available power gain $G_{Avai1} = \frac{|S_{21}|^2}{1-|S'_{22}|^2} = 0.2572$ by the insertion power gain $G_1 = |S_{21}|^2 = 0.257$ in the Friis formula can be considered negligible.

We therefore established a clear procedure for noise factor measurements over 50 Ω and validated this procedure as well as the measurement bench with an adapted attenuator whose noise factor is known in advance.

This procedure, known as the Y factor method, can now be applied to other devices to be tested as amplifiers, provided that they are output-adapted. It is therefore not suitable for a single transistor that is generally mismatched on its ouput.

This procedure can nevertheless be improved when the temperature $T_C = T_{Amb}$ is different from T_0 by replacing the equation [4.4] with the equation [4.22] for both the calibration and the measuring phase of the DUT:

$$F = \frac{ENR_{lin} - Y\left(1 - \frac{T_C}{T_0}\right)}{Y-1} \qquad [4.22]$$

If the DUT is mismatched at its output – that is, if the noise diode and the noise measurement chain remain matched at 50 Ω – at the very least a measurement of S_{22} must be done using a vectorial network analyzer, in order to calculate the available power gain of the DUT, instead of using its insertion power gain, which is different (see equations [4.19] and [4.20]). It is also necessary to insert an isolator, of bandwidth compatible with the measurement frequency/frequencies, between the output of the DUT and the measurement chain so that it sees at its input the same null reflection coefficient during calibration and measurement. The noise factor of the measuring chain (isolator + amplifier + spectrum analyzer) then remains the same during the calibration and during the measurement, which thus makes it possible to obtain coherent results.

What has not yet been explicitly stated in this chapter is that it is necessary to iterate the procedure – calibration, measurement and calculation – for each frequency point of interest and that the operation is greatly facilitated, if the spectrum analyzer and noise source are driven by a computer.

It should also be noted that the Y factor method described in this chapter is not the only one that can be used; a method known as a cold source can be used and is discussed in Exercise 4 of Chapter 6. In both cases, the reasoning is made on the available noise powers and the gains are therefore the available power gains.

The final precision concerns the setting of the resolution band of the spectrum analyzer; the following two situations should be avoided:

– choose an RBW resolution band wider than BW the bandwidth of the device under test. If BW<RBW the noise of the noise diode and that added by the device under test are only present in BW at the output of the device under test during the measurement step;

– make a measurement at a frequency f_0 close to the limits f_{Min}, f_{Max} of the bandwidth of the device under test such that $f_0 + \frac{RBW}{2} > f_{Max}$ or $f_0 - \frac{RBW}{2} < f_{Min}$.

In both cases, the term "*B*" used in the noise formulas is different for the measurement phase with the device under test and the calibration phase of the noise measurement chain. This results in false measurements.

5

Characterization in Noise

5.1. Introduction

In this chapter, we will deal with the case where we must characterize in noise any two-port, that is to say, access by measurement to knowledge, for each frequency point of interest, of:

– F_{Min} the minimum noise factor;

– R_N the equivalent noise resistance at the input;

– Y_{Sopt} the optimal source admittance for noise that contains two independent pieces of information G_{Sopt} and B_{Sopt} or Γ_{Sopt} which also contains two independent pieces of information $|\Gamma_{Sopt}|$ and $Arg(\Gamma_{Sopt})$.

As discussed in Chapter 1, to arrive at determining these four quantities, it is necessary to have a noise measurement bench which makes it possible to vary the source admittance presented at the input of the two-port. Figure 5.1 gives the synopsis of such a bench.

In such a measurement bench, we can:

– vary, with the tuner, the reflection coefficient Γ_S presented to the device under test and thus to measure the noise factor of the DUT for these different values by placing SW1 and SW2 in the red position and by measuring the noise power with a spectrum analyzer preceded by a low noise amplifier;

– measure with the vector network analyzer the S parameters of the DUT which will be needed, inter alia, to determine its gain;

– calibrate the noise measurement system and the tuner by connecting P1 and P2 directly.

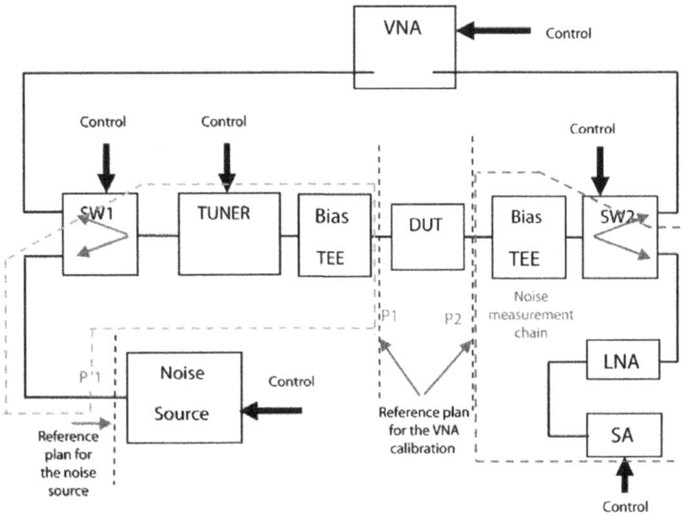

Figure 5.1. *Noise characterization bench of a two-port. For a color version of this figure, see www.iste.co.uk/fouquet/electronics.zip*

It is this calibration step that we will essentially present in this chapter because it is the one that requires the most attention for the following reasons:

– the device under test is no longer, as in Chapter 4, a device adapted at its output and has S parameters noted S_{33}, S_{34}, S_{43}, S_{44} with port 3 for the input and port 4 for the output. In addition, the reflection coefficient Γ_S presented on its input is varied and thus its output reflection coefficient $\Gamma_{OUT4} = S_{44} + \frac{S_{34} \cdot S_{43} \cdot \Gamma_S}{1 - S_{33} \cdot \Gamma_S}$ changes its value at each Γ_S presented at input. It is therefore no longer possible to consider that the measurement chain has a constant noise factor equal to F_2 (50 Ω) as in Chapter 4. It will therefore be necessary to completely characterize this measurement chain in noise to determine:

– F_{2Min} the minimum noise factor of the noise measurement chain;

– R_{2N} the equivalent noise resistance at the input of the noise measurement chain;

– Γ_{2Sopt} the optimum source reflection coefficient for noise that contains two independent pieces of information $|\Gamma_{2Sopt}|$ and $Arg(\Gamma_{2Sopt})$.

This information will make it possible to calculate $F_2(\Gamma_{OUT4})$; Γ_{OUT4} having been calculated after measurements of the parameters S of the device under test and for a given Γ_S ;

– the noise diode is no longer directly presented at the input of the device whose noise factor is to be determined but rather through the tuner responsible for modifying the reflection coefficient Γ_S presented at the input of the device under test. This cascading of the noise diode and the tuner alters the noise behavior of this set with respect to the behavior of a single diode, as used in Chapter 4.

We will explain the procedure to follow to carry out the different steps necessary for the measurements.

5.2. The tuner

5.2.1. *Tuner constitution*

Without going into the details of the construction of a tuner, it is necessary nevertheless to have some notions on the way it is realized and the characteristics that it must have. The general principle of a tuner is that of the adaptation of impedance based on the use of a stub of variable length and variable position, as shown in Figure 5.2.

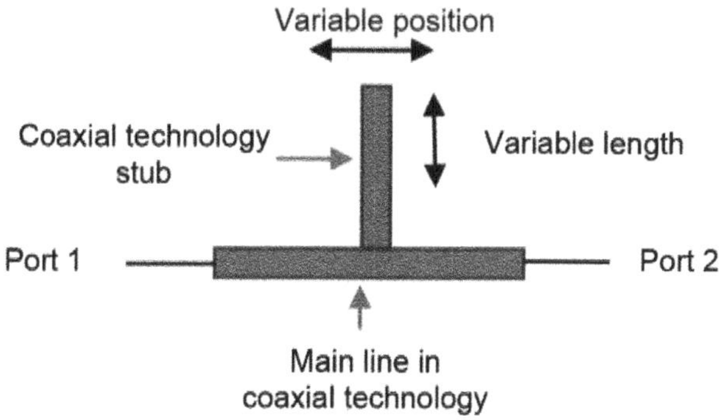

Figure 5.2. *Diagram of a tuner. For a color version of this figure, see www.iste.co.uk/fouquet/electronics.zip*

With such a device, terminated on one side by 50 Ω – the noise diode or a network analyzer port following the position of SW1 –, we can go along – (if the length of the stub and its position allow) much of Smith's chart; that is, the location of the Γ_S corresponding to impedances with a positive real part.

It is thus a passive two-port whose noise behavior can be known by using the S parameter measurement, carried out with the vector network analyzer, and Bosma's theorem already used in section 3.6 of Chapter 3.

It is advisable to use a motorized tuner – a motor for the length of the stub and a motor for the position of the stub – to be able to easily and accurately reproduce positions and stub lengths and thus the reflection coefficient Γ_S presented to the device under test.

Despite the technologies used and the care taken in the realization of a tuner, it is never free of losses. It is these losses that make it necessary to calibrate in noise the assembly consisting of the noise diode and the tuner. These losses also prevent synthesizing Γ_S with magnitudes very close to 1.

5.2.2. *Noise behavior of the noise diode + tuner assembly*

The noise diode always behaves as a resistor R_0 with a reflection coefficient $\Gamma_D = 0$ which can be set at the temperature T_C or at the temperature T_H. It is capable of supplying a noise power wave $b_{nD,C}$ at the temperature T_C or $b_{nD,H}$ at the temperature T_H. By using Bosma's theorem given by formula [3.16] for $[S] = \Gamma_D = 0$, we can calculate the expression of these noise power waves as indicated in equations [5.1a] and [5.1b]:

$$\overline{|b_{nD,C}|^2} = kT_C \cdot df \qquad [5.1a]$$

$$\overline{|b_{nD,H}|^2} = kT_H \cdot df \qquad [5.1b]$$

The tuner is a passive two-port which can be represented by its S matrix given by equation [5.2] and which is, whatever the temperature of the diode, at temperature T_C:

$$\begin{vmatrix} b_1 \\ b_2 \end{vmatrix} = \begin{vmatrix} S_{11} & S_{12} \\ S_{21} & S_{22} \end{vmatrix} \cdot \begin{vmatrix} a_1 \\ a_2 \end{vmatrix} + \begin{vmatrix} b_{n1} \\ b_{n2} \end{vmatrix} \quad [5.2]$$

This association can be represented by the flow graph of Figure 5.3.

The reflection coefficient Γ_L represents a load which will be needed to calculate the equivalent noise temperatures of the reflection coefficient Γ_S presented by the diode + tuner assembly.

The noise power waves b_{n1} and b_{n2} are the noise power waves due to the losses of the tuner as defined in section 3.6 of Chapter 3.

Seen from the load Γ_L, everything happens as if we had a source reflection coefficient Γ_S – known as measured at the network analyzer – that can take two temperatures T_{Ceq} and T_{Heq} and therefore being associated respectively with two power waves $b_{nS,C}$ and $b_{nS,C}$ according to Bosma's theorem and defined by the expressions [5.3a] and [5.3b]:

$$\left|\overline{b_{nS,C}}\right|^2 = kT_{Ceq} \cdot df \cdot (1 - |\Gamma_S|^2) \quad [5.3a]$$

$$\left|\overline{b_{nS,H}}\right|^2 = kT_{Heq} \cdot df \cdot (1 - |\Gamma_S|^2) \quad [5.3b]$$

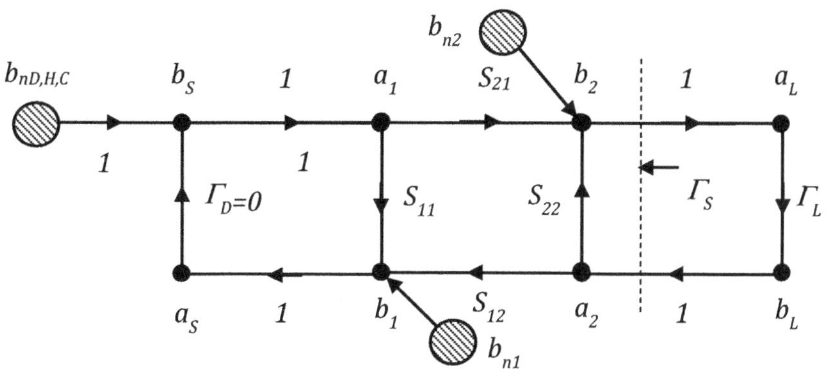

Figure 5.3. *Flow graph of the noise diode + tuner assembly*

The flow graph of Figure 5.3 can be simplified in the form given in Figure 5.4 because the reflection coefficient of the diode is $\Gamma_D = 0$.

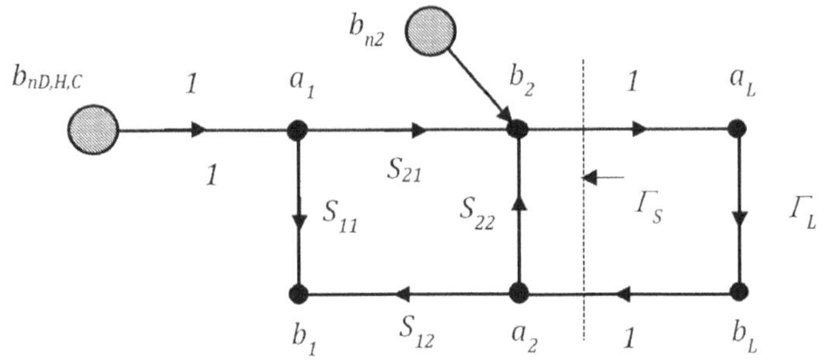

Figure 5.4. *Simplified flow graph of the noise diode + tuner assembly* ($\Gamma_D = 0$)

We will use this flow graph to express the equivalent temperatures T_{Ceq} and T_{Heq}.

The power wave incident on the load a_L can be calculated using Mason's rule and is given by equation [5.4]:

$$a_L = \frac{b_{nD,H,C} \cdot S_{21} + b_{n2}}{1 - S_{22} \cdot \Gamma_L} \qquad [5.4]$$

The corresponding available power is $P_{nAvai} = |a_L|^2 \cdot (1 - |\Gamma_L|^2)$ calculated for $\Gamma_L = S_{22}^*$ because the reflection coefficient at the output of the tuner is $\Gamma_S = S_{22}$.

For the temperature $T_{Céq}$,we can write the equality [5.5] by not omitting that $b_{nD,C}$ and $b_{n2}$are not correlated:

$$kT_{Ceq} \cdot df = \frac{\left(\left|\overline{b_{nD,C}}\right|^2 \cdot |S_{21}|^2 + |\overline{b_{n2}}|^2\right) \cdot (1 - |S_{22}|^2)}{(1 - |S_{22}|^2)^2} = \frac{\left(\left|\overline{b_{nD,C}}\right|^2 \cdot |S_{21}|^2 + |\overline{b_{n2}}|^2\right)}{1 - |S_{22}|^2} \qquad [5.5]$$

We know that $\left|\overline{b_{nD,C}}\right|^2 = kT_C \cdot df$, it remains to express $\left|\overline{b_{n2}}\right|^2$ using Bosma's theorem given by equation [3.16]. The result is given by equation [5.6]:

$$\left|\overline{b_{n2}}\right|^2 = kT_C \cdot df \cdot (1 - |S_{21}|^2 - |S_{22}|^2) \qquad [5.6]$$

Using equation [5.6] in equation [5.5], we find that $T_{Ceq} = T_C$. This result is reassuring because the noise diode and the tuner are at the same temperature T_C and synthesize a reflection coefficient Γ_S.

For temperature T_{Heq}, we can write the equality [5.7] by not omitting that $b_{nD,H}$ and b_{n2} are not correlated:

$$kT_{Heq} \cdot df = \frac{(|\overline{b_{nD,H}}|^2 \cdot |S_{21}|^2 + |\overline{b_{n2}}|^2)}{1 - |S_{22}|^2} \qquad [5.7]$$

Knowing that $|\overline{b_{nD,H}}|^2 = kT_H \cdot df$ and $|\overline{b_{n2}}|^2 = kT_C \cdot df \cdot (1 - |S_{21}|^2 - |S_{22}|^2)$, replacing in [5.7], we find equality [5.8]:

$$T_{Heq} = \frac{T_H \cdot |S_{21}|^2 + T_C \cdot (1 - |S_{21}|^2 - |S_{22}|^2)}{1 - |S_{22}|^2} = \frac{(T_H - T_C) \cdot |S_{21}|^2}{1 - |S_{22}|^2} + T_C \qquad [5.8]$$

This result would have been difficult to establish without using noise power waves and Bosma's theorem. It will be noted that the quantity $\frac{|S_{21}|^2}{1-|S_{22}|^2}$ is G_{Avai_Tun} the available power gain of the tuner when terminated at its input by the noise diode with $\Gamma_D = 0$ and that it then synthesizes at its output a reflection coefficient $\Gamma_S = \Gamma_{Out}|_{\Gamma_D=0} = S_{22} + \left.\frac{S_{21} \cdot S_{12} \cdot \Gamma_D}{1 - \Gamma_D \cdot S_{11}}\right|_{\Gamma_D=0} = S_{22}$. This available power gain therefore varies according to the Γ_S synthesized by the tuner. Note also that, from the equality [5.8], we can define, for the noise diode + tuner assembly, an $ENR_{lin,cor}$ as indicated by [5.9] and which varies according to Γ_S synthesized by the tuner:

$$ENR_{lin,cor} = \frac{T_{Heq} - T_0}{T_0} \qquad [5.9]$$

If desired or necessary, the non-adaptation of the noise diode can be included in the calculation of T_{Heq} by reasoning on the graph of Figure 5.3. This only affects the value of T_{Heq}.

5.2.3. *Tuner calibration procedure*

This tuner calibration procedure consists of:

– disconnecting Port1 from SW1 in Figure 5.1;

– performing a complete calibration of the vector network analyzer between the Port 1 and P2 planes. A LRM type calibration (Line, Reflect, Match), is perfect and not very demanding in terms of the standards required: (i) an ideal connection (i.e. you should be able to connect ports 1 and 2, so they must be of opposite polarities, male and female), (ii) male and female short circuits and (iii) male and female 50 [Ohm] loads;

– connecting Port1 of the VNA with P'1 and P1 with P2. Measuring with SW1 in red position and SW2 in blue position, for each position of the tuner (position and stub length) and for each frequency of interest, coefficients S_{21} and $S_{22}$with the network analyzer calibrated between Port1 and P2;

– depending on the ENR_{lin} of the noise diode, known for each frequency of interest, calculate, for each frequency of interest and each position of the tuner, the $ENR_{lin,cor}$ of the noise diode + tuner assembly;

– creating a calibration file that contains all this information.

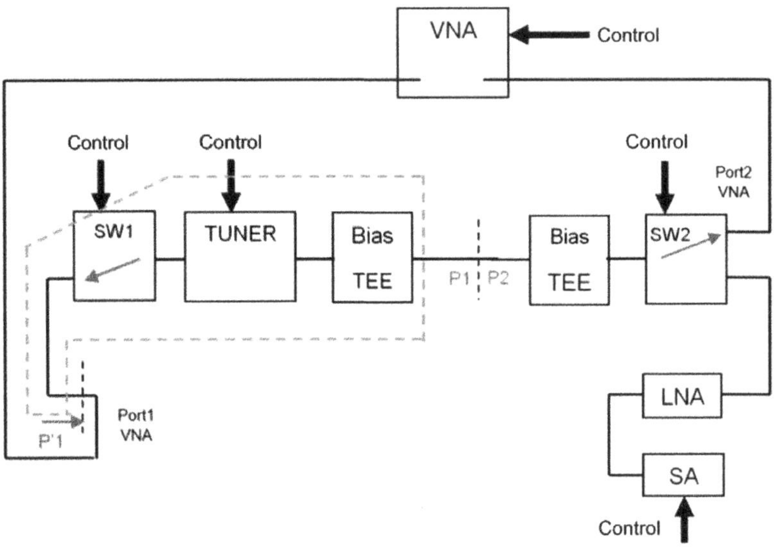

Figure 5.5. *Tuner calibration. For a color version of this figure, see www.iste.co.uk/fouquet/electronics.zip*

We can also measure the reflection coefficient of the diode Γ_D to take it into account in the calculation of T_{Heq}.

This step, unavoidable for further measurements, is usually quite tedious even with a motorized tuner and an automated bench with a computer. However, it is not necessary for each use of the bench but only periodically to overcome the problems of long-term repeatability of the correspondence between the "position and length of the stub" and "value of the synthesized Γ_S" pair at a given frequency.

Recall that the tuner part which has just been calibrated in noise consists of the entire part surrounded by the green dotted line in Figure 5.5 and must not be modified after calibration – including the cables – otherwise it will be necessary to redo this noise calibration.

5.3. Characterization in noise of the noise measurement chain

This step is, in principle, quite simple because it consists, for each frequency of interest of synthesizing quadruplets of $\Gamma_S(\Gamma_{S1}, \Gamma_{S2}, \Gamma_{S3}, \Gamma_{S4})$ of switching the noise diode from T_C to T_H, the measurements which make it possible to determine the corresponding noise factors $F_2(\Gamma_{S1})$ to $F_2(\Gamma_{S4})$, as described in section 4.4 of Chapter 4, and also using $ENR_{lin,cor}$ in the equation linking the noise factor to Y factor. From a quadruplet of $F_2(\Gamma_{Si})$, we use equation [1.16] to solve the system of equations to find F_{Min2}, R_{N2}, Γ_{Sopt2} in magnitude and in phase.

In practice, more than four reflection coefficients are used, typically a few tens of Γ_S by best covering the Smith chart, and we process them 4 by 4 to calculate, with the quadruplet i, F_{Min2-i}, R_{N2_i}, $\Gamma_{Sopt2-i}$ in magnitude and in phase. The quadruplets which give aberrant results are eliminated and with those which remain, we calculate F_{Min2}, R_{N2}, Γ_{Sopt2} in magnitude and in phase by making, for each quantity, a quadratic average.

At this point, we are able to calculate $F_2(\Gamma_{OUT1})$ for any Γ_{OUT1} present at the output of the device under test.

As shown in Figure 5.6, remember that during this step, Port1 is reconnected on SW1, switches SW1 and SW2 are both in red position and the noise diode is connected to P'1.

The noise measurement chain to be characterized is the part surrounded by the red dotted line in this Figure 5.6 and it must not be modified after characterization, including the cables, otherwise we will have to redo a complete noise characterization.

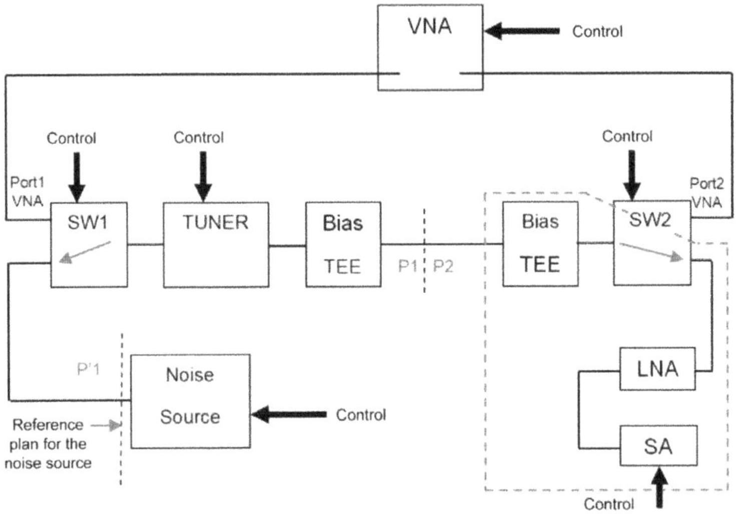

Figure 5.6. *Characterization in noise of the noise measurement chain. For a color version of this figure, see www.iste.co.uk/fouquet/electronics.zip*

5.4. Characterization in *S* parameters of the device under test

During this measurement, SW1 and SW2 are in the blue position and the device under test is inserted between P1 (input) and P2 (output). It must be equipped with connectors allowing it to be inserted between P1 and P2 without additional adapter, otherwise erroneous results will be obtained. The four *S* parameters of the device under test are then measured for each frequency of interest, after having calibrated the network analyzer between P1 and P2. In order to avoid confusion with the tuner's S parameters, those of the device under test are marked $S_{33}, S_{34}, S_{43}, S_{44}$ with port 3 for the input and port 4 for the output, as indicated in section 5.1.

From these *S* parameters, we are able to calculate $G_{Avai1}(\Gamma_S)$, the available power gain, according to Γ_S, which is needed in the Friis formula.

5.5. Noise characterization of the device under test

During this measurement, SW1 is in the red position and SW2 is in the red position. The device under test is still placed between P1 (input) and P2 (output). Using the tuner, we generate at each frequency of interest a few tens of Γ_{Si} – each known by the two parameters "position" and "length" of the stub, we do not re-measure them – and by switching the noise diode from T_C to T_H, measurements are made to determine the corresponding noise factors $F_{12}(\Gamma_{Si})$, still using $ENR_{lin,cor}$.

At each value of Γ_{Si}, we are able to calculate

– the reflection coefficient at the output of the device under test given by equation [5.10]:

$$\Gamma_{OUT4i} = S_{44} + \frac{S_{34} \cdot S_{43} \cdot \Gamma_{Si}}{1 - S_{33} \cdot \Gamma_{Si}} \qquad [5.10]$$

– the available power gain $G_{Avai1}(\Gamma_S)$ given by equation [5.11]:

$$G_{Avai1}(\Gamma_{Si}) = \frac{|S_{43}|^2 \cdot (1 - |\Gamma_{Si}|^2)}{|1 - S_{33} \cdot \Gamma_S|^2 \cdot (1 - |\Gamma_{OUT4i}|^2)} \qquad [5.11]$$

– a noise factor of the measurement chain $F_2(\Gamma_{Si})$ given by equation [5.12]:

$$F_2(\Gamma_{OUT4i}) = F_{Min2} + \frac{4 \cdot R_{N2}}{R_0} \frac{|\Gamma_{OUT4i} - \Gamma_{Sopt2}|^2}{|1 + \Gamma_{Sopt2}|^2 \cdot (1 - |\Gamma_{OUT4i}|^2)} \qquad [5.12]$$

We are therefore able to use the Friis formula to compute $F_1(\Gamma_{Si})$ given by [5.13]:

$$F_1(\Gamma_{Si}) = F_{12}(\Gamma_{Si}) - \frac{F_2(\Gamma_{OUT4i}) - 1}{G_{Avai1}(\Gamma_{Si})} \qquad [5.13]$$

With the few tens of $F_1(\Gamma_{Si})$, we apply the method described in section 5.3 to compute F_{Min1}, R_{N1}, Γ_{Sopt1} in magnitude and in phase which are the four parameters allowing us to know completely the behavior in noise of the device under test.

5.6. Validation of a noise characterization bench

5.6.1. *Introduction*

To validate such a bench, there are only two solutions:

– to characterize the same object on a reference bench and on the bench to be validated then to compare the results;

– to have a noise factor standard. The most convenient solution is to use a passive device whose noise characteristics are known in advance by applying Bosma's theorem.

This solution was chosen to validate the bench shown in Figure 1.14. Both standards are a 2.5 dB adapted attenuator and an SMA coaxial cable. The bench was calibrated and used following the procedure described in sections 5.2 to 5.5.

5.6.2. *2.5 dB adapted attenuator*

The object under test is an SMA 2.5 dB insertable attenuator, thus completely compatible with the measurement bench. This attenuator is shown in Figure 5.7.

Figure 5.7. *An SMA 2.5 dB attenuator m / f. For a color version of this figure, see www.iste.co.uk/fouquet/electronics.zip*

The noise characterization results are given in Table 5.1.

Noise parameters				
Frequency in GHz	**NF_{Min} in dB**	**R_N in Ohm**	**$\lvert\Gamma_{Sopt}\rvert$**	**$Arg(\Gamma_{Sopt})$ in °**
1000	2.42	14.72	0.001	– 91.1
1500	2.47	15.14	0.009	– 75.1
2000	2.47	15.38	0.016	– 68.5
2500	2.50	15.40	0.014	– 59.6
3000	2.52	15.62	0.012	– 66.1

Table 5.1. *2.5 dB attenuator, LRM calibration, measured values*

Knowing the *S* parameters of this attenuator and using Bosma's theorem, these same results could be calculated. They are given in Table 5.2.

Noise parameters				
Frequency in GHz	**NF_{Min} in dB**	**R_N in Ohm**	**$\|\Gamma_{Sopt}\|$**	**Arg(Γ_{Sopt}) in °**
1000	2.50	15.40	0.005	22.1
1500	2.50	15.37	0.004	– 66.5
2000	2.52	15.43	0.010	– 91.0
2500	2.52	15.58	0.016	– 71.9
3000	2.53	16.00	0.018	– 35.3

Table 5.2. *2.5 dB attenuator, LRM calibration, calculated values*

On these two tables, there is a good agreement between the measurements and the predictions made with the *S* parameters. The error on the phase of Γ_{Sopt} is not significant because the magnitude of Γ_{Sopt} is extremely weak and these points are therefore all in the center of Smith's chart.

It can be considered that the measurements with this noise bench are valid. An additional check was made by measuring the same DUT several times to validate the stability of the measurements. These results are given in Table 5.3 for a DUT which is the same 2.5 dB attenuator measured at 3 GHz.

Noise parameters				
Frequency = 3 GHz	**NF_{Min} in dB**	**R_N in Ohm**	**$\|\Gamma_{Sopt}\|$**	**Arg(Γ_{Sopt}) in °**
Measurement 1	2.52	15.62	0.012	– 66.1
Measurement 2	2.49	15.50	0.012	– 47.2
Measurement 3	2.50	15.35	0.009	– 46.0
Measurement 4	2.52	15.61	0.012	– 47.4
Measurement 5	2.48	15.57	0.013	– 43.9

Table 5.3. *2.5 dB attenuator at 3 GHz, LRM calibration, measured values*

The different measures in Table 5.3 correspond to the following cases:

– measurements 1 and 2 were made cold; that is to say right after the start of the bench and its calibration in noise. The calibration of the tuner has been done previously;

– measurements 3 and 4 were made hot, respectively two hours and four hours after measurements 1 and 2 without other noise calibrations;

– measurement 5 was made two days later without extinguishing the measurement devices but by resetting the calibration in noise of the bench's receiving chain. It is this calibration which is the most sensitive because of the variations of temperature of the low noise amplifier. It is therefore advisable to do it before each new series of measurements.

There is therefore a good reproducibility of the measurements:

– few 1/100's of dB for NF_{Min};

– about +/- 1% for the value of R_N.

5.6.3. *Coaxial cable*

A final measurement was made on a DUT with a very low noise factor. It concerns an SMA "STORM" cable of very small diameter, 10 cm in length. The losses of such a cable, connectors included, are low and therefore the associated noise factor is too; this should make it possible to see if the bench is able to correctly evaluate this type of noise factor. The results are given in Tables 5.4 and 5.5 to compare measurements and simulations.

Noise parameters				
Frequency in GHz	**NF_{Min} in dB**	**R_N in Ohm**	**$\|\Gamma_{Sopt}\|$**	**Arg(Γ_{Sopt}) in °**
1	0.13	1.07	0.12	– 19.6
2	0.25	1.54	0.06	– 100.8
3	0.31	1.89	0.03	– 93.3
4	0.34	2.79	0.13	– 16.4
5	0.45	3.02	0.18	– 59.9

Table 5.4. *STORM cable, LRM calibration, measured values*

Noise parameters				
Frequency in GHz	NF_{Min} in dB	R_N in Ohm	$\|\Gamma_{Sopt}\|$	$Arg(\Gamma_{Sopt})$ in °
1	0.25	1.52	0.01	– 1.3
2	0.31	1.76	0.08	– 98.9
3	0.36	2.08	0.01	– 132.0
4	0.38	2.88	0.13	– 16.3
5	0.42	3.12	0.17	– 49.4

Table 5.5. *STORM cable, LRM calibration, calculated values*

There is therefore a satisfactory agreement between the measurements and the predictions on the minimum noise factor (deviation less than 0.1 dB, which is lower than the uncertainty on the measurement) and the equivalent noise resistance. For the optimal reflection coefficient for noise, we are still faced with the problem of the weak magnitude which makes the value of the phase very difficult to assess.

5.7. Conclusion

In this chapter, we reviewed:

– the architecture of a noise characterization bench and the role of each element;

– the tuner reflection coefficient characterization procedure and the method for calculating the equivalent hot temperature due to the joint action of the noise diode and the tuner;

– the complete noise characterization procedure of the noise measurement system and the need perform it;

– the measurements and calculations to be performed to fully characterize a device in noise;

– a procedure to validate the proper functioning of the bench. This step is essential and, during the development of the bench presented in this chapter, it was possible to highlight that the calibration procedure initially used for the vector network analyzer – an electronic SMA f / f calibration kit – is not suitable because it caused the addition / removal of an SMA polarity adapter that was incompatible with the other bench calibration steps. The use of a calibration of the LRM network analyzer solved this problem by ensuring that the calibrations of the tuner and the noise measurement system are done exactly under the same conditions, i.e. without adding an adapter. The counterpart was the realization of a short SMA female that is not a common object.

Having this information, you just have to reproduce these different steps to build your own noise characterization bench knowing that:

– a motorized tuner is a rather specific object found in an RF or microwave laboratory only if it already has measurements in noise or load-pull type or source-pull made there. As shown in Figure 1.14, the Canadian company Focus Microwaves can provide this kind of device;

– the use of a manual tuner may bring disappointments at the time of validation of the bench because of the difficulty to reproduce positions precisely, even if it is equipped with micrometer screws. Moreover, the time spent doing the various manipulations can quickly become discouraging;

– software for controlling instrument configurations (vector network analyzer, spectrum analyzer, switches, noise diode and tuner), recovering measured information (S parameters and measured noise power), managing calibration files (tuner, noise measurement chain, $ENR_{lin,cor}$ of the diode + tuner) and calculations is essential to take full advantage of such a bench.

6

Exercises and Answers

6.1. Exercises

Exercise no. 1

Subject presentation

When transistors are used in an analog radio frequency integrated circuit, they are generally made with a single technological process which means that these transistors all have the same characteristics for the same size. The designer of the circuit then has only one possibility to adjust the transistor to his needs: play with its size. For example, in MOSFET technology, the size of the transistor is adjusted by paralleling several structures of the same elementary transistor in parallel.

The object of this exercise is to see the influence of this parallelization on the noise characteristics of the transistors obtained. To simplify this study, it will be assumed that in these parallel associations, all the elementary transistors always work with an identical polarization current.

Questions

1.1. Since the transistors are obtained by paralleling N times the structure of an elementary transistor, what is the small signal description most suitable for solving the problem?

1.2. For the elementary transistor, it is assumed that the matrix admittance and the four noise parameters $F_{Min/1}$, $R_{N/1}$, $Y_{Sopt/1}$ are known. For two of these transistors in parallel, express $F_{Min/2}$, $R_{N/2}$, $Y_{Sopt/2}$.

1.3. Generalize to *N* transistors.

1.4. We have a reference MOSFET with the following characteristics at 2 GHz: $NF_{Min.1} = 0.6dB$, $R_{N.1} = 60\ \Omega$, $\Gamma_{Sopt.1}$ = 0.9/5°. Using a calculation program, give the curve of the modulus of $\Gamma_{Sopt.N}$ for *N* varying from 0.1 to 100. The values of *N* less than 1 mean that the MOSFET considered is smaller than the reference one.

Exercise no. 2

Subject presentation

In Chapter 3, we have seen the impact of losses and therefore resistances on the noise factor. In this exercise, it is proposed to quantify the deterioration of the minimum noise factor of a transistor when it is used in an integrated circuit and that its gate connection to the previous stage is through an R_{Con} resistor that comes from, for example, the resistance of the metallic interconnection layer or how the gate is made. The transistor is a MOSFET having the following characteristics at 2 GHz (Gao 2004): 180 nm technology, 5 fingers of 16 µm, $NF_{Min} = 0.6\ dB$, $R_N = 60\ \Omega$, Γ_{Sopt} =0.9/7°. We denote *e* and *i* as the noise generators associated with this transistor and its admittance matrix [Y] is assumed to be known.

Questions

2.1. Take to describe the noise of the resistor R_{Con} an equivalent voltage generator e_R. Give its root mean square value. Say if this noise is correlated or not to that of the transistor.

2.2. Take, to describe the noise of the combination of the resistor and the transistor in cascade, an equivalent voltage generator and equivalent current generator placed at the input of this association. They are denoted as e_{Tot} and i_{Tot}. Calculate the matrix admittance $[Y_{tot}]$ of this association. Give the expressions of e_{Tot} and i_{Tot} according to e, e_R and i and express R_{Ntot}, Y_{Cortot} and G_{ntot}. The results will be validated by examining these results for $R_{Con} = 0$.

2.3. Deduce the expression of F_{MinTot} the minimum noise factor of the transistor resistance combination. Determine numerically the evolution of this quantity according to the value of this resistance.

2.4. Calculate the maximum value of R_{Con} so that NF_{MinTot} is still less than 0.7 dB.

Exercise no. 3

Subject presentation

In Chapter 4, we saw the procedure to measure the noise factor on 50 Ω of a two-port. With the bench described in this chapter, the measurements, presented in Table 6.1, were carried out at 1 GHz on a device under test.

Diode	**Calibration**	**Measurement**
Cold	– 97.2 dBm	– 92.9 dBm
Hot	– 92.4 dBm	– 83.1 dBm

Table 6.1. *Noise power measurements at 1 GHz*

Questions

3.1. Create, with Excel for example, a calculation program implementing the formulas [4.7], [4.11], [4.12], [4.14] and [4.15].

3.2. Process the measurements in Table 6.1 with the program performed and give the values in dB of the gain of the device under test and its noise figure.

3.3. The object under test is a Mini-Circuits ZHL-1010 + amplifier whose main characteristics are given in Table 6.2. Validate the measurements made and your calculation program.

Frequency (MHz)	**Gain (dB)**	**NF (dB)**
50.00	10.65	3.95
472.20	11.06	3.27
1000.00	10.72	3.8

Table 6.2. *Specifications of the ZHL-1010 + Amplifier (Mini-Circuits 2019)*

Exercise no. 4

Subject presentation

This exercise involves exploring the "cold source" method. This method can be used either to make the noise factor measurement on 50 Ω as in Chapter 4, or for the complete noise characterization as in Chapter 5. We will limit ourselves here to the measurement on 50 Ω. The measurement takes place in two stages:

– calibration of the receiver with a noise diode where we measure:

- N_{CalC};
- N_{CalH};

– the measures on the DUT where you acquire:

- N_{MesC} by noise measurement using only a cold source; for us, a 50 Ω load at T_C;
- S parameters with a vector network analyzer.

Obviously, a 2nd stage correction is used when using this method.

Questions

4.1. Starting from the definition of the noise factor which says that, for a two-port available gain G_{avai}, delivering an added noise power, measured at its output, N_A and connected to an available noise power source $k \cdot df \cdot T_0$, then $F = \frac{N_A + G_{avai} \cdot k \cdot df \cdot T_0}{G_{avai} \cdot k \cdot df \cdot T_0}$, show that the total available noise power at the output of this two-port, for a source temperature T_C, is:

$$N = G_{avai} \cdot k \cdot df(T_C + T_0(F-1))$$

4.2. The receiver is modeled for noise measurements by a noisy two-port:

– of S-matrix $[S_{Rec}] = \begin{bmatrix} S_{33} & S_{34} \\ S_{43} & S_{44} \end{bmatrix}$;

– noise factor $F_2(\Gamma_S)$;

– and loaded on its port 4 by a load which has a reflection coefficient $\Gamma_L = 0$.

This receiver is powered on its port 3 by a noise source of reflection coefficient Γ_S and available power $P_{SAvai,n}$, as indicated in Figure 6.1. Recall that for noise, it is necessary to account for available powers. Express the noise power collected in Γ_L due to the noise source as a function of $P_{SAvai,n}$ the available power of that source and the S parameters of the receiver.

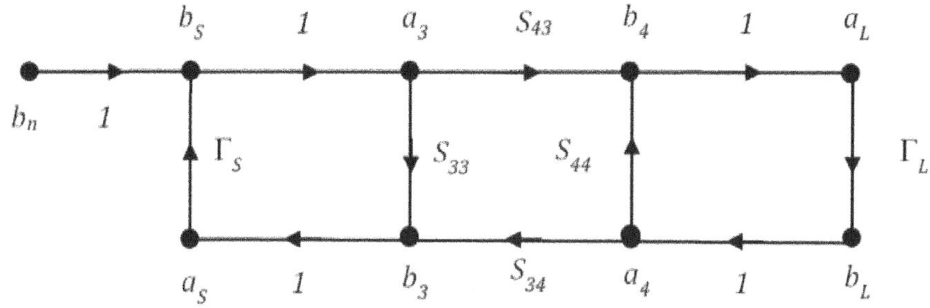

Figure 6.1. *Modeling of the receiver*

State what the terms $|S_{43}|^2$and $\frac{1-|\Gamma_S|^2}{|1-\Gamma_S \cdot S_{33}|^2}$ physically represent.

4.3. N_{A2} denotes the available noise power added by the receiver and measured at its output and Γ_D the reflection coefficient of the noise diode assumed to be independent of the equivalent noise temperature of the diode. Using the wording found in the previous question, give the literal expressions of the two noise powers measured during the calibration of the receiver respectively denoted N_{CalH} and N_{CalC}.

4.4. Show that we can deduce from these two measurements the quantities $F_2(\Gamma_D)$ and $k \cdot df \cdot |S_{43}|^2$ after measurements of Γ_D and S_{33} with a vector network analyzer.

4.5. The DUT is inserted between a 50 Ω load at temperature T_c, and the receiver. N_{A1} is the available noise power added by the DUT and measured at its output, G_{Avai1} its available power gain and S_{22} the reflection coefficient seen on its output. Give, with these notations, the expression of the noise power N_{MesC} measured by the receiver.

4.6. Say what are the known quantities and unknown quantities after the three noise measurements. Using the answer to question 4.1, deduce the expression of the noise factor of the DUT + receiver set denoted $F_{12}(\Gamma = 0)$.

4.7. Deduce the use that one will make of the measurement of parameters S of the DUT with the vector network analyzer.

4.8. State what to do if S_{22} is different from Γ_D.

4.9. Give the expression of $F_1(\Gamma = 0)$ assuming $F_2(S_{22})$ is known.

Exercise no. 5

Subject presentation

This exercise proposes to demonstrate the Friis formula using the formalism of noise power waves. Reading of Appendix 4 is essential to carry out the exercise. We model

– the device under test by its S matrix and its noise power waves:

$$[S_{DUT}] = \begin{bmatrix} S_{11} & S_{12} \\ S_{21} & S_{22} \end{bmatrix}$$

b_{n1}, b_{n2}

– the noise measurement chain by its S matrix and its noise power waves:

$$[S_{Réc}] = \begin{bmatrix} S_{33} & S_{34} \\ S_{43} & S_{44} \end{bmatrix}$$

b_{n3}, b_{n4}

To solve this problem whose solution is known, we will calculate the S parameters of the DUT set in cascade with the measurement chain and the noise power waves associated with the two ports of this set, its noise factor $F_{12}(\Gamma_S)$ and identify each of its terms to show $F_1(\Gamma_S)$, $F_2(\Gamma_{OUT})$ and $G_{Avail}(\Gamma_S)$. The flow graph of the set is given in Figure 6.2.

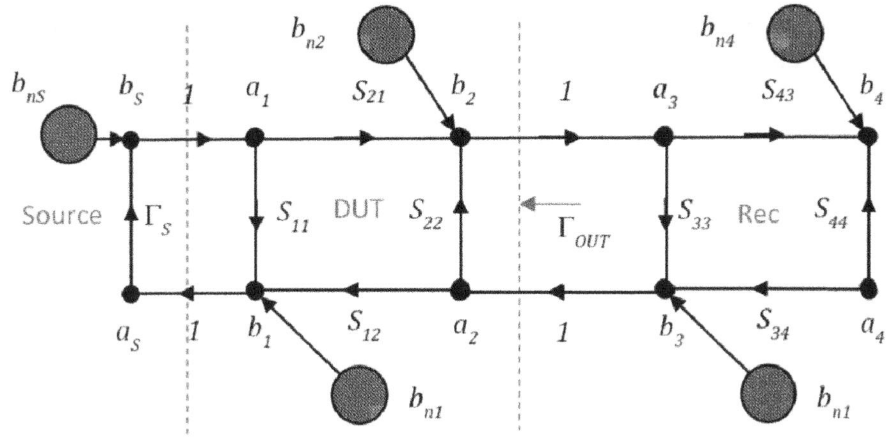

Figure 6.2. *Flow graph of DUT and receiver. For a color version of this figure, see www.iste.co.uk/fouquet/electronics.zip*

Questions

5.1. Calculate S_{11tot} and S_{21tot} the reflection coefficient on port 1 and the transmission coefficient between ports 1 and 4 of the DUT + receiver set using the Mason's rule.

5.2. Calculate b_{n1tot} and b_{n2tot} the equivalent power waves on ports 1 and 4 of the DUT + receiver set using the Mason's rule.

5.3. Express $F_{12}(\Gamma_S)$ according to S_{11tot}, S_{21tot}, b_{n1tot} and b_{n2tot}. What are the values of the correlation terms between the noise power waves of the DUT and those of the receiver? Separate $F_{12}(\Gamma_S)$ accordingly in two terms.

5.4. Express the component in b_{n1} and b_{n2} of $F_{12}(\Gamma_S)$. Simplify your expression and say what this term is.

5.5. Give the expression of Γ_{OUT} the reflection coefficient seen by the receiver on its input when the DUT is terminated at its input by Γ_S.

5.6. Express $F_2(\Gamma_{OUT}) - 1$ and $G_{Avai1}(\Gamma_S)$ and express the ratio of these two quantities.

5.7. Express the contribution of b_{n4} to $F_{12}(\Gamma_S)$ and identify with the result of question 5.6.

5.8. Express the contribution of b_{n3} to $F_{12}(\Gamma_S)$ and identify with the result of question 5.6.

5.9. Conclude on $F_{12}(\Gamma_S)$ expressed using noise power waves.

Exercise no. 6

Subject presentation

In Chapter 1, we defined the relations that can exist between the noise generators e and i in the representation of Figure 1.11 and the noise parameters of a two-port, which are F_{Min}, R_N and Y_{Sopt}. In fact, the quantities of interest for noise factor calculations are contained in the correlation matrix of the noise generators. For this representation of Figure 1.11, we need to know C_C defined by:

$$[C_C] = \begin{bmatrix} \overline{|e|^2} & \overline{e.\iota^*} \\ \overline{e^*.\iota} & \overline{|\iota|^2} \end{bmatrix} = 4kT \cdot df \cdot \begin{bmatrix} c_{C11} & c_{C12} \\ c_{C21} & c_{C22} \end{bmatrix}$$

Questions

6.1. Express c_{C11} according to R_N.

6.2. Express c_{C22} according to R_N.and Y_{Sopt}.

6.3. Express c_{C21} and c_{C12} according to R_N, F_{eMin} and Y_{Sopt}.

Exercise 7

Subject presentation

In Chapter 1, we have used two noise generators e and i to describe the noise of the two-port in relation to the matrix admittance and the representation of Figure 1.11. Now, the most natural representation when using the admittance matrix is rather that of Figure 1.10. In this exercise, we will establish the relationships between these two representations because the representation of Figure 1.10 is of interest when it comes to representing the noise in the intrinsic part of a MOSFET transistor as shown in Figure 6.3. Indeed, for the intrinsic part of a MOSFET, the noise is often expressed from two noise current generators i_d and i_g which the origin can be physically explained and which are obviously partially correlated.

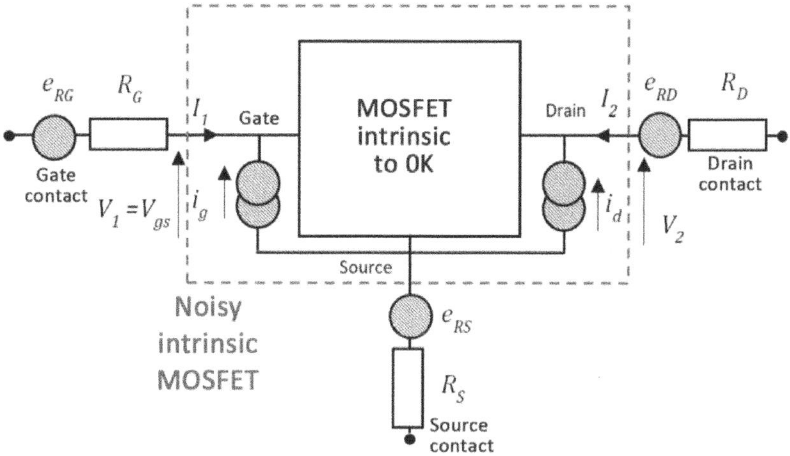

Figure 6.3. *Representation of a noisy MOSFET. For a color version of this figure, see www.iste.co.uk/fouquet/electronics.zip*

The intrinsic part of the MOSFET is suitably described for small signal operation by the admittance matrix Y_{Mos_int}.

$$[Y_{Mos\_int}] = \begin{bmatrix} Y_{11} & Y_{12} \\ Y_{21} & Y_{22} \end{bmatrix}$$

$$= \begin{bmatrix} \dfrac{j\omega C_{gs}}{1 + j\omega R_i C_{gs}} + j\omega C_{gd} & -j\omega C_{gd} \\ g_m - j\omega C_{gd} & g_{ds} + j\omega(C_{ds} + C_{gd}) \end{bmatrix}$$

For the intrinsic MOSFET, noise models (Asgaramet 2004, Gao 2004) were developed on the basis of Van der Ziel's work (Van der Ziel 1963) giving the expressions of $\overline{|\iota_d|^2}$, $\overline{|\iota_g|}^2$ and $\overline{\iota_g^* \cdot \iota_d}$. These models have in common, despite details of notations, to describe these three quantities in the following form:

– $\overline{|\iota_d|^2} = 4kT \cdot df \cdot g_d$ with g_d independent of the frequency and referring through a coefficient either to g_m the transconductance of the

MOSFET or to g_{d0} the Drain / Source conductance at $V_{DS} = 0$. The generator i_d represents the thermal noise of the channel;

– $\overline{|\imath_{gtot}|^2} = \overline{|\imath_{g0}|^2} + \overline{|\imath_g|^2} = \overline{|\imath_{g0}|^2} + 4kT \cdot df \cdot \frac{(\omega \cdot C_0)^2}{g_m}$ with C_0 independent of the frequency and referring through a coefficient to C_{gs} the gate / source capacitance of the MOSFET. We see that $\overline{|\imath_g|^2}$ increases in f^2. In Van der Ziel's initial model, the noise generator i_{g0} is shot noise;

– $\overline{\imath_g \cdot \imath_d^*} = j \cdot 4kT \cdot df \cdot \omega \cdot C_0$ with C_0 independent of the frequency and referring through a coefficient to C_{gs} the gate / source capacitance of the MOSFET. We see that $\overline{\imath_g \cdot \imath_d^*}$ grows toward *f*.

– Van der Ziel's explanation for the form of $\overline{|\imath_g|^2}$ and $\overline{\imath_g \cdot \imath_d^*}$ is as follows (Van der Ziel 1963). He argues that the total noise current i_{gtot} has two components:

– a component i_{g0} due to the carriers (electrons and holes) which, randomly, arrive on the gate or leave it. This component is uncorrelated to i_d because of different origin;

– a component i_g due to the distributed nature of the channel along the source / drain axis. Indeed, the noise current i_d induces potential differences along this axis through the elementary resistances of the channel represented in the small signal model by R_i. This results by capacitive coupling in a noise current i_g in the gate. This current i_g is completely correlated with i_d.

In practice, we find that $i_g > i_{g0}$ thus $i_{gtot} \# i_g$.

It is clear that this model refers to gate / channel junction field effect transistors (JFET or MESFET) because of the explanation given for the nature of i_{g0} (shot noise) which exists only for junctions. In a MOSFET transistor, the gate is isolated and this shot noise does not exist. Consequently, the portion of i_{gtot} uncorrelated to i_d is of a different nature. In this exercise, we will see to what extent the Van der Ziel model can be applied to a MOSFET transistor.

Questions

7.1. Express the noise factor of the two-port of Figure 1.10 according to its admittance matrix, the source admittance Y_S connected to its input and the two noise generators i_1 and i_2. For this, the power collected in a non-noisy load Y_L will be expressed as a function of V_2 and then, using Mason's rules, V_2 will be expressed as a function of i_{nS}, i_1 and i_2.

7.2. Establish the transitional relationships between the two noise generators e and i used to describe the two-port noise in Figure 1.11 and the two noise generators i_1 and i_2 used to describe the noise of the two-port in Figure 1.10.

7.3. Check that the noise factor obtained in question 7.1 which shows Y_{11} and Y_{21} is that given by equation [1.11].

7.4. Using the answer to question 7.1, express $\overline{|i_2|^2}$ assuming $g_m \gg j\omega C_{gd}$ and conclude on the compatibility with the form proposed by Van der Ziel.

7.5. Using the answer to question 7.1, express $\overline{i_1 \cdot i_2^*}$ assuming $g_m \gg \omega C_{gd}$. Deduce the condition on Y_{Cor} so that the expression proposed by Van der Ziel is valid. Check whether the Van der Ziel model on the form of $\overline{i_g \cdot i_d^*}$ is valid for a MOSFET having the following characteristics at 2 GHz (Gao 2004): 180 nm technology, 5 fingers of 16 μm, $NF_{Min} = 0.37\,dB$, $R_N = 60\Omega$, $\Gamma_{Sopt} = 0.93/5°$, $g_m = 57mS$, $C_{gs} = 125fF$, $C_{gd} = 41fF$, $R_i = 2\Omega$. To answer the question, $Re(Y_{11})$ will be expressed.

7.6. Using the answer to question 7.1, express $\overline{|i_1|^2}$ assuming $g_m \gg j\omega C_{gd}$ and conclude on compatibility with the form proposed by Van der Ziel.

6.2. Solutions

Solution to exercise no. 1

1.1 The simplest way is to work in matrix admittance because during a paralleling, admittance parameters are added, as well as currents.

1.2. For each transistor i = 1 or 2, we can use the formulation given by equation [2.4]:

$$\begin{vmatrix} I_{1i} \\ I_{2i} \end{vmatrix} = \begin{bmatrix} Y_{11} & Y_{12} \\ Y_{21} & Y_{22} \end{bmatrix} \cdot \begin{vmatrix} V_{1i} \\ V_{2i} \end{vmatrix} + \begin{bmatrix} Y_{11} & -1 \\ Y_{21} & 0 \end{bmatrix} \cdot \begin{vmatrix} e_i \\ i_i \end{vmatrix} \qquad [2.4]$$

For two identical transistors in parallel, we can write:

$$\begin{vmatrix} I_{1tot} \\ I_{2tot} \end{vmatrix} = \begin{bmatrix} 2Y_{11} & 2Y_{12} \\ 2Y_{21} & 2Y_{22} \end{bmatrix} \cdot \begin{vmatrix} V_{1tot} \\ V_{2tot} \end{vmatrix} + \begin{bmatrix} 2Y_{11} & -1 \\ 2Y_{21} & 0 \end{bmatrix} \cdot \begin{vmatrix} e_{tot} \\ i_{tot} \end{vmatrix}$$

By letting in the equivalent circuit $V_{1i} = V_{2i} = V_{1tot} = V_{2tot} = 0$, we can calculate the contributions of e_i and i_i to I_{1tot} and I_{2tot} then identify $e_{tot}\ and\ i_{tot}$. We find:

$$e_{tot} = \frac{(e_1 + e_2)}{2} \qquad i_{tot} = i_1 + i_2$$

From there and without forgetting that the noise generators of one transistor are not correlated with those of the second transistor, we can compute $R_{N\cdot 2} = \frac{1}{2} \cdot R_{N\cdot 1}$, $Y_{Cor\cdot 2} = \frac{\overline{i_{tot} \cdot e_{tot}^*}}{\overline{e_{tot}^2}} = 2 \cdot Y_{Cor\cdot 1}$, $i_{ntot} = i_{n1} + i_{n2}$ and $G_{n\cdot 2} = 2G_{n\cdot 1}$. We then obtain:

$$G_{Sopt\cdot 2} = \sqrt{G_{Cor\cdot 2}{}^2 + \frac{G_{n\cdot 2}}{R_{N\cdot 2}}} = 2 \cdot G_{Sopt\cdot 1} \text{and } B_{Sopt\cdot 2} = 2 \cdot B_{Sopt\cdot 1}.$$

We can therefore conclude that $F_{Min\cdot 2} = 1 + 2 \cdot R_{N\cdot 2} \cdot (G_{Sopt\cdot 2} + G_{Cor.2}) = F_{Min.1}$.

1.3. By generalizing to N identical transistors – by iteration of the previous calculation – we obtain:

$$F_{Min\cdot N} = F_{Min\cdot 1}$$

$$R_{N\cdot N} = \frac{1}{N} \cdot R_{N\cdot 1}$$

$$G_{Sopt\cdot N} = N \cdot G_{Sopt\cdot 1}$$

$$B_{Sopt\cdot N} = N \cdot B_{Sopt\cdot 1}$$

As interesting information, we find that it is the technological process responsible for realizing the transistor which fixes its minimum noise factor, and not the size of the transistor. On the other hand, what we can do is adapt the size of the transistor, via the parameter N, to minimize the quantity $\frac{\frac{R_{N\cdot1}}{N}\cdot\left|Y_S-N\cdot Y_{Sopt\cdot1}\right|^2}{G_S}$. This is done for a given impedance source $Y_S = G_S + jB_S$ and for a given technology; that is, for $R_{N\cdot1}$ *and* $Y_{Sopt\cdot1}$ data.

1.4. Using $\Gamma_{Sopt\cdot1}$, we can calculate $y_{Sopt\cdot1} = 0{,}0527 - 0{,}0435j$ and deduce the values of $y_{Sopt.i}$ to recalculate the $\Gamma_{Sopt.i}$ for *i* ranging from 0.1 to 100. The recurrence equation is written:

$$\Gamma_{Sopt.i} = \frac{\frac{1-i}{1+i} + \Gamma_{Sopt\cdot1}}{1 - \frac{1-i}{1+i}\cdot\Gamma_{Sopt\cdot1}}$$

This recurrence equation is not easy to interpret because of $\Gamma_{Sopt\cdot1}$ which is a complex number; we must therefore compute numerically the $\Gamma_{Sopt.i}$. The curve of the magnitude of $\Gamma_{Sopt.i}$ is given below. Note that this magnitude is minimum for i = 15. The corresponding reflection coefficient is $\Gamma_{Sopt.i}$= 0.36/92.3°. This makes it possible to obtain a figure of noise of 0.74 dB for a 50 Ω source. This result is of course only relative to the transistor used.

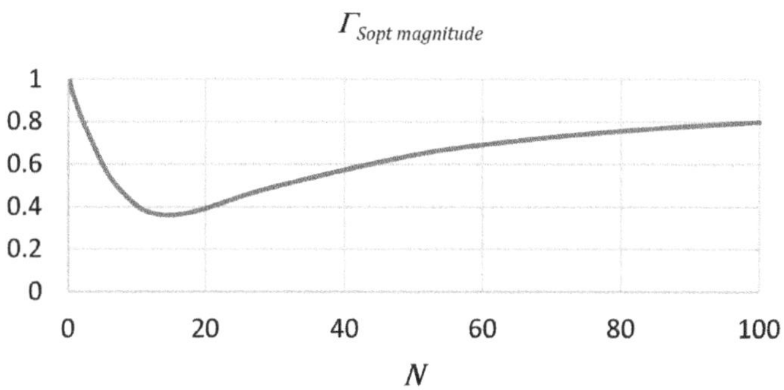

Figure 6.4. *Magnitude of the optimum source reflection coefficient for noise as a function of the number of MOSFETs*

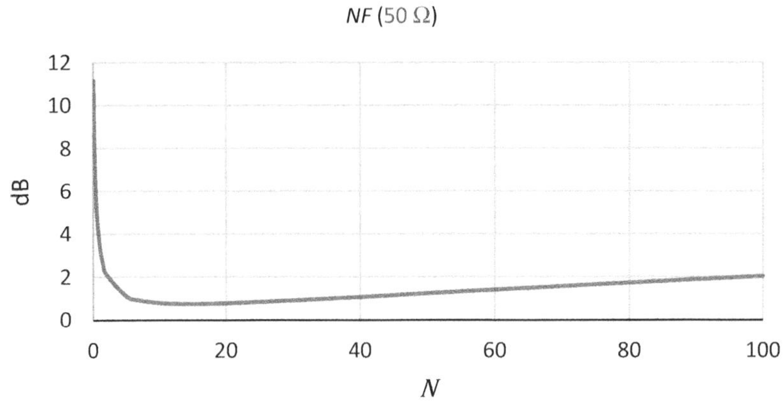

Figure 6.5. *50 Ω Noise Figure as a function of the number of MOSFETs*

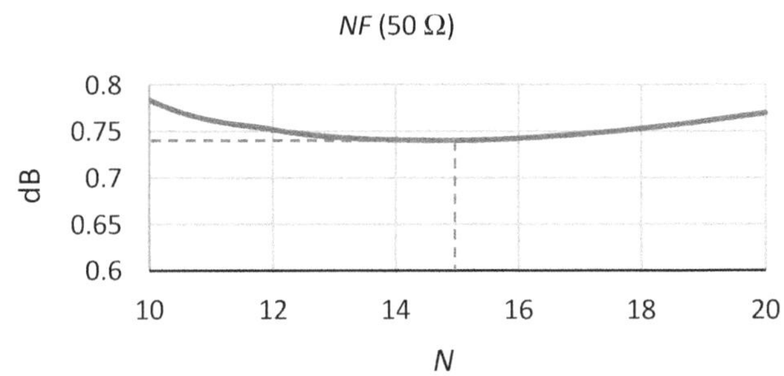

Figure 6.6. *Minimum 50 Ω Noise Figure*

The reasoning for this exercise does not take into account the losses brought by the elements used to connect together the elementary transistors. This point is addressed in Exercise no. 2.

Solution to exercise no. 2

2.1. We have $\overline{e_R}^2 = 4kT \cdot df \cdot \mathrm{R}_{Con}$ and this generator is not correlated to those of the transistor.

2.2. For the resistance + transistor assembly, we can write:

$$\begin{vmatrix} I_{1tot} \\ I_{2tot} \end{vmatrix} = \begin{bmatrix} Y_{11tot} & Y_{12tot} \\ Y_{21tot} & Y_{22tot} \end{bmatrix} \cdot \begin{vmatrix} V_{1tot} \\ V_{2tot} \end{vmatrix} + \begin{bmatrix} Y_{11tot} & -1 \\ Y_{21tot} & 0 \end{bmatrix} \cdot \begin{vmatrix} e_{tot} \\ i_{tot} \end{vmatrix}$$

where e_{tot} and i_{tot} are the equivalent generators to e, i and e_R.

In the matrix $[Y_{tot}]$, we just need Y_{11tot} and Y_{21tot} which are calculated at $V_{2tot} = e = i = e_R = 0$. We obtain $Y_{11tot} = \frac{Y_{11}}{1+R_{Con} \cdot Y_{11}}$ and $Y_{21tot} = \frac{Y_{21}}{1+R_{Con} \cdot Y_{11}}$.

Then, we have to calculate the contributions of e_R, e and i to I_{1tot} and I_{2tot} at $V_{1tot} = V_{2tot} = 0$.

For e, we get $I_{1tot} = Y_{11tot} \cdot e$ and $I_{2tot} = Y_{21tot} \cdot e$.

For e_R, we get $I_{1tot} = Y_{11tot} \cdot e_R$ and $I_{2tot} = Y_{21tot} \cdot e_R$.

For i, we get $I_{1tot} = -\frac{1}{1+R_{Con} \cdot Y_{11}} \cdot i$ and $I_{2tot} = Y_{21tot} \cdot R_{Con} \cdot i$.

We deduce that $e_{tot} = e + e_R + R_{Con} \cdot i$ and $i_{tot} = i$.

We can then express $R_{Ntot} = R_{Con} + R_{Con}^2 \cdot G_n + R_N \cdot |1 + Y_{Cor} \cdot R_{Con}|^2$. We check that $R_{Ntot} = R_N$ for $R_{Con} = 0$. We can write R_{Ntot} in the form:

$$R_{Ntot} = R_N + R_{Con}\left[1 + 2R_N G_{Cor} + R_N \cdot R_{Con} \cdot \left|Y_{Sopt}\right|^2\right]$$

It is clear that $R_{Ntot} > R_N$.

To calculate Y_{Cortot}, we use the equation $Y_{Cortot} = \frac{\overline{i_{tot} \cdot e_{tot}^*}}{\overline{e_{tot}^2}}$ and we obtain:

$$Y_{Cortot} = \frac{\overline{(i_n + Y_{Cor} \cdot e) \cdot (e^*(1 + Y_{Cor}^* \cdot R_{Con}) + R_{Con} \cdot i_n^* + e_R^*)}}{\overline{e_{tot}^2}}$$

In the numerator of Y_{Cortot}, only the terms in $\overline{i_n \cdot i_n^*}$ and in $\overline{e \cdot e^*}$ are non-zero. We obtain:

$$Y_{Cortot} = \frac{R_{Con} \cdot G_n + R_N \cdot (Y_{Cor} + |Y_{Cor}|^2 \cdot R_{Con})}{R_{Ntot}}$$

$$= \frac{R_{Con} \cdot G_n + R_N \cdot (Y_{Cor} + |Y_{Cor}|^2 \cdot R_{Con})}{R_{Con} + R_{Con}^2 \cdot G_n + R_N \cdot |1 + Y_{Cor} \cdot R_{Con}|^2}$$

We check that $Y_{Cortot} = Y_{Cor}$ for $R_{Con} = 0$.

Y_{Cortot} can be rewritten as:

$$Y_{Cortot} = \frac{R_N \cdot (Y_{Cor} + |Y_{Sopt}|^2 \cdot R_{Con})}{R_N + R_{Con}\left[1 + 2R_N G_{Cor} + R_N \cdot R_{Con} \cdot |Y_{Sopt}|^2\right]}$$

We thus obtain:

$$G_{Cortot} = \frac{R_N \cdot (G_{Cor} + |Y_{Sopt}|^2 \cdot R_{Con})}{R_N + R_{Con}\left[1 + 2R_N G_{Cor} + R_N \cdot R_{Con} \cdot |Y_{Sopt}|^2\right]}$$

$$= \frac{R_N \cdot (G_{Cor} + |Y_{Sopt}|^2 \cdot R_{Con})}{R_{Ntot}}$$

$$B_{Cortot} = \frac{R_N \cdot B_{Cor}}{R_N + R_{Con}\left[1 + 2R_N G_{Cor} + R_N \cdot R_{Con} \cdot |Y_{Sopt}|^2\right]} = \frac{R_N \cdot B_{Cor}}{R_{Ntot}}$$

It is difficult to conclude on G_{Cortot} with respect to G_{Cor}, but this is not a problem because in the minimum noise factor, the important term is the product "noise equivalent resistance at the input by real part of the correlation admittance". For the transistor with resistance, we have a product that is equal to:

$$G_{Cortot} \cdot R_{Ntot} = R_N \cdot \left(G_{Cor} + |Y_{Sopt}|^2 \cdot R_{Con}\right) > R_N \cdot G_{Cor} \qquad [6.1]$$

We are therefore moving towards an increase in the minimum noise factor in the presence of R_{Con} unless a decrease in $R_{Ntot} \cdot G_{Sopttot}$ compensates for the increase on $R_{Ntot} \cdot G_{Cortot}$. To calculate $G_{Sopttot}$, we need G_{ntot}.

To express this quantity G_{ntot}, we can attempt to compute $i_{ntot} = i_{tot} - Y_{Cortot} \cdot e_{tot}$ but this may be complicated. The simplest is to use a peculiarity of the circuit: $i_{tot} = i$.

We can write:

$$i_{tot} = i_{ntot} + Y_{Cortot} \cdot e_{tot} = i = i_n + Y_{Cor} \cdot e$$

We therefore immediately obtain:

$$G_{ntot} + |Y_{Cortot}|^2 \cdot R_{Ntot} = G_n + |Y_{Cor}|^2 \cdot R_N \qquad [6.2]$$

Let: $G_{ntot} = G_n + |Y_{Cor}|^2 \cdot R_N - |Y_{Cortot}|^2 \cdot R_{Ntot}$. This gives the value of R_{Ntot} but does not allow to conclude on $G_{Sopttot}$.

2.3. To be able to advance, we must rewrite equation [6.2] in the form of equation [6.3]:

$$|Y_{Sopttot}|^2 \cdot R_{Ntot} = |Y_{Sopt}|^2 \cdot R_N \qquad [6.3]$$

By rewriting equation [6.3] in the form of equation [6.4], we can obtain $G_{Sopttot}$ in the form given by equation [6.5]:

$$R_{Ntot} \cdot \left(G_{Sopttot}{}^2 + B_{Sopttot}{}^2\right) = R_N \cdot \left(G_{Sopt}{}^2 + B_{Sopt}{}^2\right) \qquad [6.4]$$

with:

$$B_{Sopttot} = \frac{R_N}{R_{Ntot}} \cdot B_{Sopt}$$

$$G_{Sopttot} = \sqrt{\frac{R_N}{R_{Ntot}}} \cdot \sqrt{\left(G_{Sopt}{}^2 + B_{Sopt}{}^2 (1 - \frac{R_N}{R_{Ntot}})\right)} \qquad [6.5]$$

Let: $R_{Ntot} \cdot G_{Sopttot} = \sqrt{R_N \cdot R_{Ntot}} \cdot \sqrt{\left(G_{Sopt}{}^2 + B_{Sopt}{}^2 (1 - \frac{R_N}{R_{Ntot}})\right)}$. [6.6]

As: $R_{Ntot} > R_N$ thus $\sqrt{R_N \cdot R_{Ntot}} > R_N$ and $\sqrt{\left(G_{Sopt}{}^2 + B_{Sopt}{}^2 (1 - \frac{R_N}{R_{Ntot}})\right)} > G_{Sopt}$ therefore: $R_{Ntot} \cdot G_{Sopttot} > R_N \cdot G_{Sopt}$.

This asserts that $F_{eMintot} > F_{eMin}$.

To calculate $F_{eMintot}$, the simplest way is to use equations [6.1] and [6.6] because $F_{eMintot} = 2.R_{Ntot}.(G_{Sopttot} + G_{Cortot})$ and to obtain equation [6.7]

$$F_{Mintot} = 1 + 2.\left[\sqrt{R_N \cdot R_{Ntot}} \cdot \sqrt{\left(G_{Sopt}{}^2 + B_{Sopt}{}^2\left(1 - \frac{R_N}{R_{Ntot}}\right)\right)} + \frac{R_N \cdot (G_{Cor} + |Y_{Sopt}|^2 \cdot R_{Con})}{R_{Ntot}}\right] \quad [6.7]$$

with:

$$R_{Ntot} = R_N + R_{Con}\left[1 + 2R_N G_{Cor} + R_N \cdot R_{Con} \cdot |Y_{Sopt}|^2\right]$$

The result of F_{Mintot} calculation is given in Table 6.3 for MOSFET 1 whose characteristics are the following at 2 GHz: $NF_{Min} = 0.6$ dB, $R_N = 60\ \Omega$, $\Gamma_{Sopt} = 0.9/7°$.

$R_{Con}(\Omega)$	0	5	7.5	10	12.5	15	17.5	20
r_{Con}	0	0.10	0.15	0.20	0.25	0.30	0.35	0.40
NF_{Mintot}	0.6	0.65	0.67	0.70	0.72	0.74	0.76	0.78

Table 6.3. *Figure of minimum noise according to the connection resistance*

2.4. It can be seen in Table 6.3 that to stay below 0.7 dB, the connection resistance must remain below 10 Ω with this transistor. For 50 Ω of connection resistance, a noise figure of 0.98 dB is obtained, of which 0.38 dB is attributable to this resistance.

Figure 6.7 gives the appearance of NF_{Mintot} of MOSFET 1 as a function of the normalized value of R_{Con} at 50 Ω. In this figure, we see the minimum noise figure of another MOSFET 2 very comparable to the first. Its characteristics at 2 GHz are: 180 nm technology, 20 fingers of 5 μm, $NF_{Min} = 0.6\ dB$, $R_N = 60\ \Omega$, $\Gamma_{Sopt} = 0.9/14.3°$ (Asgaram *et al.* 2004). The increase of NF_{Mintot} is much faster and one reaches 0.7 dB with only 3.5 Ω.

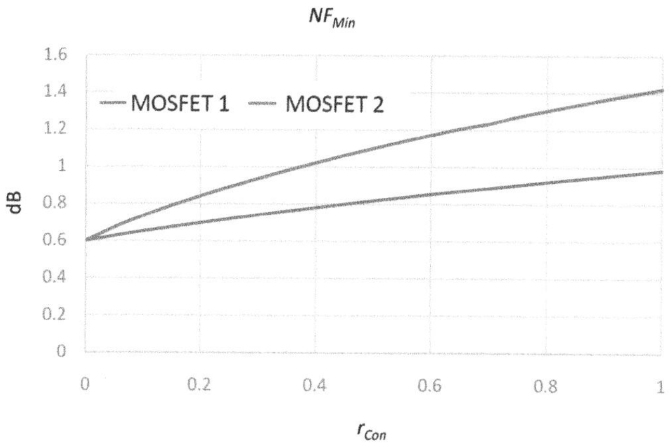

Figure 6.7. *Evolution of the minimum noise figure as a function of the normalized value of the connection resistance for two types of MOSFETs*

We therefore see the benefit of having a very low gate resistance and care interconnections in an RFCMOS integrated circuit to not lose the noise qualities of the transistors used, especially when it comes to achieving a low noise amplifier.

To lower the gate resistance, there is a known technique which consists in producing the transistor with several fingers as shown in Figure 6.8. All these MOSFETs have, at identical polarization, the same transconductance g_m, but the more the number of fingers N increases, the more the gate resistance decreases, N^2 as a first approximation, and the noise due to this resistance decreases.

It will be noted that the lower the phase of the optimum reflection coefficient for the noise at the operating frequency, the smaller the transistor NF_{Mintot} even with significant values of resistance R_{Con}. This corresponds to optimal source admittance for noise such that g_{Sopt} *and* b_{Sopt} is very small compared to 1; this is possible if the working frequency f is small compared to $f_T = \frac{g_m}{2\pi C_{gg}}$ where $C_{gg} = C_{gs} + C_{gd}$ and g_m is the transconductance of the MOSFET (Asgaram *et al.* 2004).

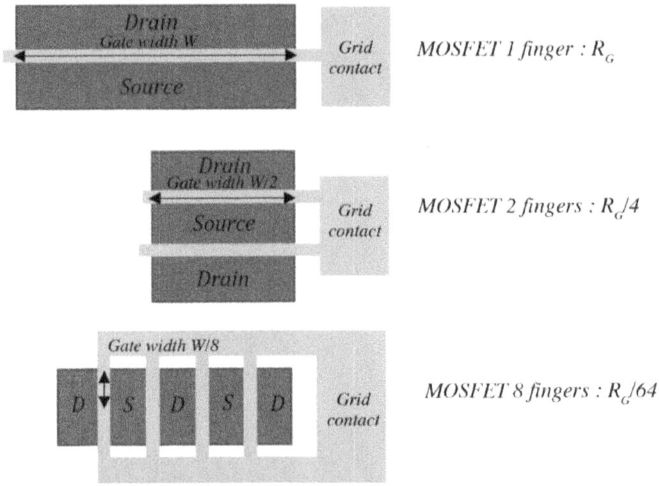

Figure 6.8. *Examples of topologies of MOSFETs with multiple gate fingers*

Solutions to exercise 3

To use this calculation routine, at the measurement frequency, enter the following 5 values:

Noise diode's ENR in dB: 14.74

	The measured noise power without DUT for Tc in dBm:	-97.2
CAL	The measured noise power without DUT for Th in dBm:	-92.4
	The measured noise power with DUT for Tc in dBm:	-92.9
MES	The measured noise power with DUT for Th in dBm:	-83.1

	ENR(dB)	ENR(lin)			
	14.74	29.78516429			
	CAL			MES	
Tc	-97.2	1.90546E-10	Tc	-92.9	5.12861E-10
Th	-92.4	5.7544E-10	Th	-83.1	4.89779E-09
Y		3.01995172			9.54992586
Y-1		2.01995172			8.54992586
	F2(lin)	14.74548327		F12(lin)	3.483675155
	F2(dB)	11.68659011		F12(dB)	5.420376512
	G1(lin)	11.39256092			
	G1(dB)	10.56621359			
	F1(lin)	2.277143685			
	F1(dB)	3.573904349			

Figure 6.9. *Extraction results in Excel. For a color version of this figure, see www.iste.co.uk/fouquet/electronics.zip*

3.1. The worksheet already used in Chapter 4 is used here to process the four measured powers and find the DUT noise figure and its gain. We obtain G1=10.57 dB and F1=3.6 dB.

3.2. This corresponds well to the characteristics of the ZHL-1010 + as shown in Table 6.4.

Frequency (MHz)	Gain (dB)	NF (dB)
50.00	10.65	3.95
472.20	11.06	3.27
1000.00	10.72	3.8

Table 6.4. *Gain and Noise Figure of the amplifier Mini-Circuits ZHL-1010 + (manufacturer data)*

Solution to exercise no. 4

4.1. By definition, we have $F = \frac{N_{A+}G_{Avai}\cdot k\cdot df\cdot T_0}{G_{Avai}\cdot k\cdot df\cdot T_0}$. We can write $N_A = (F-1).G_{Avai}\cdot k\cdot df\cdot T_0$. By measuring the total noise power at the output of the two-port at T_C, we obtain:

$$N = k\cdot df\cdot T_C\cdot G_{Avai} + N_A = k\cdot df.G_{Avai}\cdot (T_C + T_0\cdot (F-1))$$

This expression will be used in question 4.6 to write on a cold source measurement:

$$F = \frac{N}{k\cdot df\cdot G_{Avai}\cdot T_0} - \frac{T_C - T_0}{T_0}$$

4.2. We have $P_L = |a_L|^2 = \frac{|\overline{b_n}|^2\cdot |S_{43}|^2}{|1-S_{33}\cdot \Gamma_S|^2}$ and $P_{SAvai,n} = \frac{|\overline{b_n}|^2}{1-|\Gamma_S|^2}$.

Therefore:

$$P_L = P_{SAvai,n}\cdot\frac{|S_{43}|^2\cdot (1-|\Gamma_S|^2)}{|1-S_{33}\cdot\Gamma_S|^2}$$

The term $|S_{43}|^2$ corresponds to the gain of the receiver and the term $MM(\Gamma_S) = \frac{(1-|\Gamma_S|^2)}{|1-S_{33}\cdot\Gamma_S|^2}$ takes into account the mismatch between the source Γ_S and the receiver through S_{33}.

4.3. Taking into account the gain of the receiver and the mismatch between the diode and the receiver, we can write:

$$N_{CalC} = k \cdot T_C \cdot df \cdot |S_{43}|^2 \cdot MM(\Gamma_D) + N_{A2}$$

$$N_{CalH} = k \cdot T_H \cdot df \cdot |S_{43}|^2 \cdot MM(\Gamma_D) + N_{A2}$$

with: $MM(\Gamma_D) = \frac{(1-|\Gamma_D|^2)}{|1-S_{33}\cdot\Gamma_D|^2}$.

4.4. Using equation [4.4], we obtain:

$$F_2(\Gamma_D) = \frac{\frac{T_H - T_0}{T_0} - Y_{Cal} \cdot \frac{T_C - T_0}{T_0}}{Y_{Cal} - 1}$$

with: $Y_{Cal} = \frac{N_{CalH}}{N_{CalC}}$.

On the other hand, we have also: $\frac{N_{CalH}-N_{CalC}}{T_H-T_C} = k \cdot df \cdot |S_{43}|^2 \cdot MM(\Gamma_D)$. The term $MM(\Gamma_D)$ is obtained by measuring Γ_D and S_{33} with the vector network analyzer. So we know the term $k \cdot df \cdot |S_{43}|^2$ also called the "kGB" of the receiver where G is the gain of the receiver and B its bandwidth.

4.5. We have:

$$N_{MesC} = k \cdot T_C \cdot df \cdot G_{Avai1} \cdot |S_{43}|^2 . MM(S_{22}) \\ + N_{A1} \cdot |S_{43}|^2 \cdot MM(S_{22}) + N_{A2}$$

4.6. In this expression, only G_{Avai1} and the term $MM(S_{22})$ are unknown.

We obtain using answer 4.1:

$$F_{12}(\Gamma = 0) = \frac{N_{MesC}}{k \cdot df \cdot G_{Avai1} \cdot |S_{43}|^2 \cdot MM(S_{22}) \cdot T_0} - \frac{T_C - T_0}{T_0}$$

4.7. To determine G_{Avai1} and $MM(S_{22})$, we use the measurement of the S parameters of the DUT with the vector network analyzer. This time all the terms are known.

4.8. Using the Friis formula, one can write: $F_{12}(\Gamma = 0) = F_1(\Gamma = 0) + \frac{F_2(S_{22})-1}{G_{Avai1}(\Gamma=0)}$.

If Γ_D and S_{22} are different, this means that $F_2(\Gamma_D)$ determined in question 4.4 is not necessarily equal to $F_2(S_{22})$ which is needed to calculate $F_1(\Gamma = 0)$. The only solution is to do a complete noise characterization of the receiver as described in chapter 5 to calculate $F_2(S_{22})$.

4.9. Finally, if $F_2(S_{22})$ is known, we have:

$$F_1(\Gamma = 0) = \frac{N_{MesC}}{k \cdot df \cdot G_{Avai1} \cdot |S_{43}|^2 \cdot MM(S_{22}) \cdot T_0} - \frac{F_2(S_{22}) - 1}{G_{Avai1}(\Gamma = 0)} - \frac{T_C - T_0}{T_0}$$

In the general case, we have:

$$F_1(\Gamma_S) = \frac{N_{MesC}}{k \cdot df \cdot |S_{43}|^2 \cdot G_{Avai1} \cdot MM(S'_{22}) \cdot T_0} - \frac{F_2(S'_{22}) - 1}{G_{Avai1}(\Gamma_S)} - \frac{T_C - T_0}{T_0}$$

with: $S'_{22} = S_{22} + \frac{S_{12}S_{21}\Gamma_S}{1-S_{11}\Gamma_S}$, $(S'_{22}) = \frac{(1-|S'_{22}|^2)}{|1-S_{33} \cdot S'_{22}|^2}$, $G_{Avai1} = \frac{|S_{21}|^2 \cdot (1-|\Gamma_S|^2)}{|1-S_{11} \cdot \Gamma_S|^2 \cdot \left(1-|S'_{22}|^2\right)}$ and the term $k \cdot df \cdot |S_{43}|^2$ is determined during calibration or noise characterization of the receiver.

To conclude on this cold source method, we can say that it has several advantages:

– it makes it possible to perform a complete noise characterization of the receiver by using once the noise diode at T_H; all other measurements are done with a cold source at T_C using a tuner;

– the characterization of the DUT is done only with cold sources synthesized with the load 50 Ω and the tuner;

– it takes into account natively a correction of impedance mismatches; which is not the case of the method with the factor Y;

– this method is one that can be easily integrated into a vector network analyzer.

Solution to exercise no. 5

5.1. By Mason, we get:

$$S_{11tot} = S_{11} + \frac{S_{12} \cdot S_{21} \cdot S_{33}}{1 - S_{33} \cdot S_{22}}$$

$$S_{21tot} = \frac{S_{21} \cdot S_{43}}{1 - S_{33} \cdot S_{22}}$$

5.2. By Mason, we get:

$$b_{n1tot} = b_{n1} + \frac{S_{12} \cdot S_{33}}{1 - S_{33} \cdot S_{22}} \cdot b_{n2} + \frac{S_{12}}{1 - S_{33} \cdot S_{22}} \cdot b_{n3}$$

$$b_{n2tot} = \frac{S_{43}}{1 - S_{33} \cdot S_{22}} . b_{n2} + \frac{S_{22} . S_{43}}{1 - S_{33} \cdot S_{22}} \cdot b_{n3} + b_{n4}$$

5.3. Using [A4.6], we can write:

$$F_{12}(\Gamma_S) = 1 + \frac{\overline{\left|\Gamma_S \cdot S_{21tot} \cdot b_{n1tot} + (1 - \Gamma_S \cdot S_{11tot}) \cdot b_{n2tot}\right|^2}}{|S_{21tot}|^2 \cdot \overline{|b_{nS}|^2}}$$

with: $\overline{|b_{nS}|^2} = kT \cdot df \cdot (1 - |\Gamma_S|^2)$.

As the noise power waves in the DUT are not correlated with those of the measurement chain, we can write:

$$F_{12}(\Gamma_S) = 1 + \frac{\overline{\left|\Gamma_S \cdot S_{21tot} \cdot b_{n1tot}|_{b_{n3}=b_{n4}=0} + (1 - \Gamma_S \cdot S_{11tot}) \cdot b_{n2tot}|_{b_{n3}=b_{n4}=0}\right|^2}}{|S_{21tot}|^2 \cdot \overline{|b_{nS}|^2}}$$

$$+ \frac{\overline{\left|\Gamma_S \cdot S_{21tot} \cdot b_{n1tot}|_{b_{n1}=b_{n2}=0} + (1 - \Gamma_S \cdot S_{11tot}) \cdot b_{n2tot}|_{b_{n1}=b_{n2}=0}\right|^2}}{|S_{21tot}|^2 \cdot \overline{|b_{nS}|^2}}$$

with:

$$b_{n1tot}|_{b_{n3}=b_{n4}=0} = b_{n1} + \frac{S_{12} \cdot S_{33}}{1 - S_{33} \cdot S_{22}} \cdot b_{n2}$$

$$b_{n2tot}|_{b_{n3}=b_{n4}=0} = \frac{S_{43}}{1 - S_{33} \cdot S_{22}} \cdot b_{n2}$$

$$b_{n1tot}|_{b_{n1}=b_{n2}=0} = \frac{S_{12}}{1 - S_{33} \cdot S_{22}} \cdot b_{n3}$$

$$b_{n2tot}|_{b_{n1}=b_{n2}=0} = \frac{S_{22} \cdot S_{43}}{1 - S_{33} \cdot S_{22}} \cdot b_{n3} + b_{n4}$$

It remains only to identify the term:

$$\frac{\overline{\left|\Gamma_S \cdot S_{21tot} \cdot b_{n1tot}|_{b_{n3}=b_{n4}=0} + (1 - \Gamma_S \cdot S_{11tot}) \cdot b_{n2tot}|_{b_{n3}=b_{n4}=0}\right|^2}}{|S_{21tot}|^2 \cdot \overline{|b_{nS}|^2}}$$

which must be equal to $F_1(\Gamma_S) - 1$ and the term:

$$\frac{\overline{\left|\Gamma_S \cdot S_{21tot} \cdot b_{n1tot}|_{b_{n1}=b_{n2}=0} + (1 - \Gamma_S \cdot S_{11tot}) \cdot b_{n2tot}|_{b_{n1}=b_{n2}=0}\right|^2}}{|S_{21tot}|^2 \cdot \overline{|b_{nS}|^2}}$$

which must be equal to $\frac{F_2(\Gamma_{OUT})-1}{G_{Avai1}(\Gamma_S)}$.

5.4. To identify the first term which is the component in b_{n1}, b_{n2} of $F_{12}(\Gamma_S)$, we calculate:

$$\frac{\Gamma_S \cdot S_{21tot} \cdot b_{n1tot}|_{b_{n3}=b_{n4}=0} + (1 - \Gamma_S \cdot S_{11tot}) \cdot b_{n2tot}|_{b_{n3}=b_{n4}=0}}{S_{21tot}}$$

$$= \frac{\Gamma_S \cdot S_{21}}{S_{21}} \cdot b_{n1} + \frac{\Gamma_S \cdot S_{21tot}}{S_{21tot}} \cdot \frac{S_{12} \cdot S_{33}}{1 - S_{33} \cdot S_{22}} \cdot b_{n2}$$

$$+ \frac{(1 - \Gamma_S \cdot S_{11tot})}{S_{21tot}} \cdot \frac{S_{43}}{1 - S_{33} \cdot S_{22}} . b_{n2}$$

$$= \frac{\Gamma_S \cdot S_{21}}{S_{21}} \cdot b_{n1} + \frac{\Gamma_S \cdot S_{12} \cdot S_{33}}{1 - S_{33} \cdot S_{22}} \cdot b_{n2} + \frac{(1 - \Gamma_S \cdot S_{11tot})}{S_{21}} \cdot b_{n2}$$

$$= \frac{\Gamma_S \cdot S_{21}}{S_{21}} \cdot b_{n1} + \frac{\Gamma_S \cdot S_{12} \cdot S_{33} \cdot S_{21}}{S_{21} \cdot (1 - S_{33} \cdot S_{22})} \cdot b_{n2}$$

$$+ \frac{(1 - S_{33} \cdot S_{22} + \Gamma_S \cdot (S_{11} \cdot S_{22} - S_{12} \cdot S_{21}) \cdot S_{33}}{S_{21} \cdot (1 - S_{33} \cdot S_{22})} \cdot b_{n2}$$

$$= \frac{\Gamma_S \cdot S_{21}}{S_{21}} \cdot b_{n1} + \frac{(1 - S_{33} \cdot S_{22}) \cdot (1 - \Gamma_S \cdot S_{11})}{S_{21} \cdot (1 - S_{33} \cdot S_{22})} \cdot b_{n2}$$

$$= \frac{\Gamma_S \cdot S_{21}}{S_{21}} \cdot b_{n1} + \frac{(1 - \Gamma_S \cdot S_{11})}{S_{21}} \cdot b_{n2}$$

This term, once divided by b_{nS} and taken in squared modulus, corresponds to $F_1(\Gamma_S) - 1$ by identification with equation [A4.6].

5.5. By Mason's rule, we find that:

$$\Gamma_{OUT} = S_{22} + \frac{S_{12} \cdot S_{21} \cdot \Gamma_S}{1 - S_{11} \cdot \Gamma_S} = \frac{S_{22} + \Gamma_S \cdot (S_{12} \cdot S_{21} - S_{11} \cdot S_{22})}{1 - S_{11} \cdot \Gamma_S}$$

5.6. To express $F_2(\Gamma_{OUT}) - 1$, we use equation [A4.6] again, i.e.:

$$F_2(\Gamma_{OUT}) - 1 = \frac{\left|\overline{\Gamma_{OUT} \cdot S_{43} \cdot b_{n3} + (1 - \Gamma_{OUT} \cdot S_{33}) \cdot b_{n4}}\right|^2}{|S_{43}|^2 \cdot |\overline{b_{nS'}}|^2}$$

with:

$$|\overline{b_{nS'}}|^2 = kT \cdot df \cdot (1 - |\Gamma_{OUT}|^2)$$

The available power gain of the DUT is:

$$G_{Avai1}(\Gamma_S) = \frac{|S_{21}|^2 \cdot (1 - |\Gamma_S|^2)}{(1 - |\Gamma_{OUT}|^2) \cdot |1 - \Gamma_S \cdot S_{11}|^2}$$

Therefore:

$$\frac{F_2(\Gamma_{OUT}) - 1}{G_{Avai1}(\Gamma_S)}$$

$$= \frac{\left|\overline{\Gamma_{OUT} \cdot S_{43} \cdot b_{n3} + (1 - \Gamma_{OUT} \cdot S_{33}) \cdot b_{n4}}\right|^2 \cdot (1 - |\Gamma_{OUT}|^2) \cdot |1 - \Gamma_S \cdot S_{11}|^2}{|S_{43}|^2 \cdot |S_{21}|^2 \cdot (1 - |\Gamma_S|^2) \cdot kT \cdot df \cdot (1 - |\Gamma_{OUT}|^2)}$$

$$\frac{F_2(\Gamma_{OUT})-1}{G_{Avail}(\Gamma_S)} = \frac{\left|\overline{\Gamma_{OUT}\cdot S_{43}\cdot b_{n3}+(1-\Gamma_{OUT}\cdot S_{33})\cdot b_{n4}}\right|^2\cdot|1-\Gamma_S\cdot S_{11}|^2}{|S_{43}|^2\cdot|S_{21}|^2\cdot(1-|\Gamma_S|^2)\cdot kT\cdot df}$$

$$\frac{F_2(\Gamma_{OUT})-1}{G_{Avail}(\Gamma_S)} = \frac{\left|\overline{\Gamma_{OUT}\cdot S_{43}\cdot b_{n3}+(1-\Gamma_{OUT}\cdot S_{33})\cdot b_{n4}}\right|^2\cdot|1-\Gamma_S\cdot S_{11}|^2}{|S_{43}|^2\cdot|S_{21}|^2\cdot|\overline{b_{nS}}|^2}$$

$$\frac{F_2(\Gamma_{OUT})-1}{G_{Avail}(\Gamma_S)} = \frac{\left|\overline{\Gamma_{OUT}\cdot S_{43}\cdot(1-\Gamma_S\cdot S_{11})\cdot b_{n3}+(1-\Gamma_{OUT}\cdot S_{33})\cdot(1-\Gamma_S\cdot S_{11})\cdot b_{n4}}\right|^2}{|S_{43}|^2\cdot|S_{21}|^2\cdot|\overline{b_{nS}}|^2}$$

As a consequence, the contribution of b_{n3} to $\frac{F_2(\Gamma_{OUT})-1}{G_{Avail}(\Gamma_S)}$ must be:

$$\frac{\Gamma_{OUT}\cdot S_{43}\cdot(1-\Gamma_S\cdot S_{11})}{S_{43}\cdot S_{21}} = \frac{\Gamma_{OUT}\cdot(1-\Gamma_S\cdot S_{11})}{S_{21}}$$

That of b_{n4} must be:

$$\frac{(1-\Gamma_{OUT}\cdot S_{33})\cdot(1-\Gamma_S\cdot S_{11})}{S_{43}\cdot S_{21}}$$

5.7. We take again the expression of $F_{12}(\Gamma_S)\cdot|b_{nS}|^2$ according to b_{n3}, b_{n4}:

$$\frac{\left|\overline{\Gamma_S\cdot S_{21tot}\cdot b_{n1tot}|_{b_{n1}=b_{n2}=0}+(1-\Gamma_S\cdot S_{11tot})\cdot b_{n2tot}|_{b_{n1}=b_{n2}=0}}\right|^2}{|S_{21tot}|^2}$$

The contribution of b_{n4} is therefore:

$$\frac{(1-\Gamma_S\cdot S_{11tot})}{S_{21tot}} = \frac{1-\Gamma_S\cdot\dfrac{(S_{11}\cdot(1-S_{33}\cdot S_{22})+S_{33}\cdot S_{21}\cdot S_{12})}{1-S_{33}\cdot S_{22}}}{\dfrac{S_{43}\cdot S_{21}}{1-S_{33}\cdot S_{22}}}$$

$$\frac{(1-\Gamma_S\cdot S_{11tot})}{S_{21tot}}=\frac{1-\Gamma_S\cdot S_{11}-S_{33}(S_{22}+\Gamma_S\cdot(-S_{11}\cdot S_{22}+S_{21}\cdot S_{12})}{S_{43}\cdot S_{21}}$$

$$\frac{(1-\Gamma_S\cdot S_{11tot})}{S_{21tot}}=\frac{(1-\Gamma_S\cdot S_{11})\cdot(1-S_{33}\cdot\frac{S_{22}+(-S_{11}\cdot S_{22}+S_{21}\cdot S_{12})\cdot\Gamma_S}{1-\Gamma_S\cdot S_{11}})}{S_{43}\cdot S_{21}}$$

$$\frac{(1-\Gamma_S\cdot S_{11tot})}{S_{21tot}}=\frac{(1-\Gamma_S\cdot S_{11})\cdot\left(1-S_{33}\cdot\left(S_{22}+\frac{S_{21}\cdot S_{12}\cdot\Gamma_S}{1-\Gamma_S\cdot S_{11}}\right)\right)}{S_{43}\cdot S_{21}}$$

$$\frac{(1-\Gamma_S\cdot S_{11tot})}{S_{21tot}}=\frac{(1-\Gamma_S\cdot S_{11})\cdot(1-S_{33}\cdot\Gamma_{OUT})}{S_{43}\cdot S_{21}}$$

So, the contribution of b_{n4} to the term:

$$\frac{\left|\Gamma_S\cdot S_{21tot}\cdot b_{n1tot}|_{b_{n1}=b_{n2}=0}+(1-\Gamma_S\cdot S_{11tot})\cdot b_{n2tot}|_{b_{n1}=b_{n2}=0}\right|^2}{|S_{21tot}|^2\cdot|\overline{b_{nS}}|^2}$$

is that given by b_{n4} at $\frac{F_2(\Gamma_{OUT})-1}{G_{Avai1}(\Gamma_S)}$.

5.8. We take again the expression $F_{12}(\Gamma_S)\cdot|b_{nS}|^2$ according to b_{n3}, b_{n4}:

$$\frac{\left|\Gamma_S\cdot S_{21tot}\cdot b_{n1tot}|_{b_{n1}=b_{n2}=0}+(1-\Gamma_S\cdot S_{11tot})\cdot b_{n2tot}|_{b_{n1}=b_{n2}=0}\right|^2}{|S_{21tot}|^2}$$

The contribution of b_{n3} is therefore:

$$\frac{\Gamma_S\cdot S_{21tot}}{S_{21tot}}\cdot\frac{S_{12}}{1-S_{33}\cdot S_{22}}+\frac{1-\Gamma_S\cdot S_{11tot}}{S_{21tot}}\cdot\frac{S_{22}\cdot S_{43}}{1-S_{33}\cdot S_{22}}$$

$$\frac{\Gamma_S\cdot S_{12}}{1-S_{33}\cdot S_{22}}+\frac{(1-\Gamma_S\cdot S_{11tot})\cdot S_{22}}{S_{21}}$$

$$\frac{\Gamma_S\cdot S_{12}}{1-S_{33}\cdot S_{22}}+\frac{\left(1-\Gamma_S\cdot(S_{11}+\frac{S_{12}\cdot S_{21}\cdot S_{33}}{1-S_{33}\cdot S_{22}})\right)\cdot S_{22}}{S_{21}}$$

$$\frac{\Gamma_S\cdot S_{12}}{1-S_{33}\cdot S_{22}}+\frac{(1-S_{33}\cdot S_{22}-\Gamma_S\cdot(S_{11}+S_{33}\cdot(-S_{11}.S_{22}+S_{12}.S_{21}))).S_{22}}{S_{21}\cdot(1-S_{33}\cdot S_{22})}$$

$$\frac{S_{22} - S_{33} \cdot S_{22} \cdot S_{22} - \Gamma_S \cdot (S_{11} \cdot S_{22} - S_{12} \cdot S_{21}) + \Gamma_S \cdot S_{33} \cdot S_{22}(-S_{11} \cdot S_{22} + S_{12} \cdot S_{21})}{S_{21} \cdot (1 - S_{33} \cdot S_{22})}$$

$$\frac{(1 - S_{33} \cdot S_{22}) \cdot (S_{22} - \Gamma_S \cdot (S_{11} \cdot S_{22} - S_{12} \cdot S_{21}))}{S_{21} \cdot (1 - S_{33} \cdot S_{22})} = \frac{S_{22} - \Gamma_S \cdot (S_{11} \cdot S_{22} - S_{12} \cdot S_{21})}{S_{21}}$$

but:

$$\Gamma_{OUT} = \frac{S_{22} + \Gamma_S \cdot (S_{12} \cdot S_{21} - S_{11} \cdot S_{22})}{1 - S_{11} \cdot \Gamma_S}$$

therefore, the contribution of b_{n3} is equal to:

$$\frac{\Gamma_{OUT} \cdot (1 - S_{11} \cdot \Gamma_S)}{S_{21}}$$

This is the expected value because it appears in $\frac{F_2(\Gamma_{OUT})-1}{G_{Avai1}(\Gamma_S)}$.

5.9. We find the form of the Friis formula with the noise power waves but the proof is more complicated than the one proposed in Chapter 2, in particular for the term $\frac{F_2(\Gamma_{OUT})-1}{G_{Avai1}(\Gamma_S)}$ that was not directly identified, but only through the contributions of the noise power waves b_{n3} and b_{n4} and without moving to the squared mean squared value to simplify the calculations.

Solution to exercise no. 6

To answer the questions, we take again the notations of section 1.4.3 given by equations [1.12], [1.14] and [1.15]:

$$|\bar{e}|^2 = 4 \cdot kT_0 . R_N df, i = i_n + Y_{Cor} \cdot e, Y_{Cor} = G_{Cor} + jB_{Cor}, |\bar{i_n}|^2$$
$$= 4 \cdot kT_0 \cdot G_n df, Y_{Sopt} = G_{Sopt} + jB_{Sopt} = \sqrt{G_{Cor}^2 + \frac{G_n}{R_N}} - jB_{Cor}$$

And $F_{eMin} = F_{Min} - 1 = 2 \cdot R_N \cdot (G_{Sopt} + G_{Cor})$

6.1. We obtain directly $c_{C11} = R_N$.

6.2· To express c_{C22}, we compute $|\bar{\imath}|^2 = |\bar{\imath_n}|^2 + |Y_{Cor}|^2 \cdot |\bar{e}|^2 = 4kT \cdot df \cdot \left[\frac{G_n}{R_N} + G_{Cor}^2 + B_{Cor}^2\right] \cdot R_N$. By replacing $\frac{G_n}{R_N} + G_{Cor}^2$ by G_{Sopt}^2 and B_{Cor}^2 by B_{Sopt}^2, we get $c_{C22} = \left|Y_{Sopt}\right|^2 \cdot R_N$.

6.3. To express c_{C12}, we calculate $\overline{e \cdot \imath^*} = Y_{Cor}^* \cdot |\bar{e}|^2 = 4kT \cdot df \cdot (G_{Cor} - jB_{Cor},) \cdot R_N$. Now, we have $B_{Sopt} = -B_{Cor}$et $G_{Cor} = \frac{F_{eMin}}{2} - G_{Sopt}$. The result is: $c_{C12} = \frac{F_{eMin}}{2} - Y_{Sopt}^* \cdot R_N$. We then obtain: $c_{C21} = \frac{F_{eMin}}{2} - Y_{Sopt} \cdot R_N = c_{C12}^*$.

This correlation matrix of the generators e and i is denoted C_C because in general this noise representation is used with the chain matrix to describe the two-port, even if we have widely used it with the matrix admittance.

Solution to exercise no. 7

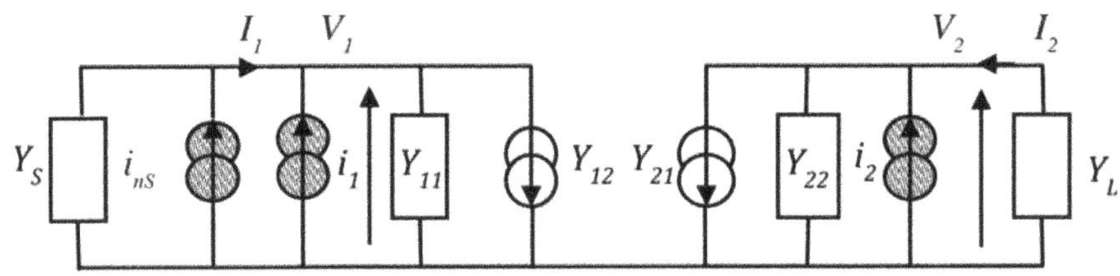

Figure 6.10. *Two-port known by its admittance matrix loaded at the input by a source of admittance YS*

7.1. With the notations of Figure 6.10, we can write: $P_L = -Re(V_2 \cdot I_2^*) = |V_2|^2 \cdot Re(Y_L)$.

The noise factor of the two-port is:

$$F(Y_S) = 1 + \frac{|V_2|_{i_{nS}=0}^2}{|V_2|_{i_1=i_2=0}^2}$$

It only remains to express V_2 according to i_{nS}, i_1 and i_2 using Mason's rule, for example. We obtain:

$$V_2 = \frac{-Y_{21} \cdot i_{nS}}{(Y_S + Y_{11}) \cdot (Y_L + Y_{22}) - Y_{12} \cdot Y_{21}}$$

$$V_2 = \frac{-Y_{21} \cdot i_1}{(Y_S + Y_{11}) \cdot (Y_L + Y_{22}) - Y_{12} \cdot Y_{21}}$$

$$V_2 = \frac{(Y_{11} + Y_S) \cdot i_2}{(Y_S + Y_{11}) \cdot (Y_L + Y_{22}) - Y_{12} \cdot Y_{21}}$$

So, the noise factor is:

$$F(Y_S) = 1 + \frac{\overline{\left| i_1 - \frac{(Y_{11} + Y_S)}{Y_{21}} \cdot i_2 \right|^2}}{\overline{|i_{nS}|^2}}$$

7.2. For the representation of Figure 1.11, we have:

$$\begin{vmatrix} I_1 \\ I_2 \end{vmatrix} = \begin{bmatrix} Y_{11} & Y_{12} \\ Y_{21} & Y_{22} \end{bmatrix} \cdot \begin{vmatrix} V_1 \\ V_2 \end{vmatrix} + \begin{bmatrix} Y_{11} & -1 \\ Y_{21} & 0 \end{bmatrix} \cdot \begin{vmatrix} e \\ i \end{vmatrix}$$

For the representation of Figure 1.10, we have:

$$\begin{vmatrix} I_1 \\ I_2 \end{vmatrix} = \begin{bmatrix} Y_{11} & Y_{12} \\ Y_{21} & Y_{22} \end{bmatrix} \cdot \begin{vmatrix} V_1 \\ V_2 \end{vmatrix} - \begin{vmatrix} i_1 \\ i_2 \end{vmatrix}$$

By identification, we obtain the following relationships:

$$i_1 = -Y_{11} \cdot e + i$$

$$i_2 = -Y_{21} \cdot e$$

7.3. We use these relations of passage to verify the expression of $F(Y_S)$:

$$\overline{\left| i_1 - \frac{(Y_{11} + Y_S)}{Y_{21}} \cdot i_2 \right|^2} = \overline{\left| -Y_{11} \cdot e + i + \frac{(Y_{11} + Y_S)}{Y_{21}} \cdot Y_{21} \cdot e \right|^2} = \overline{|i + Y_S \cdot e|^2}$$

Therefore, the expression of $F(Y_S)$ found in question 7.1 is well compatible with expression [1.11], given the transient relationships found in question 7.2. So do not worry about the presence of Y_{21} and Y_{11} in the expression found in question 7.1.

7.4. Using the passage equation, we obtain:

$$\overline{|\iota_2|^2} = |Y_{21}|^2 \cdot \overline{|e|^2} = 4kT \cdot df \cdot |Y_{21}|^2 \cdot R_N$$

This expression is well consistent with Van der Ziel's model $\overline{|\iota_d|^2} = 4kT \cdot df \cdot g_d$ provided that the term $|Y_{21}|^2 \cdot R_N \# g_m^2 \cdot R_N$ does not vary over the frequency range where the model is used. In Chen *et al.* (2001), there are experimental results for the intrinsic parts of MOSFETs of different gate lengths that prove the validity of the model. These intrinsic characteristics were obtained after de-embedding of the S parameters and noise measurements. These results are shown in Figure 6.11.

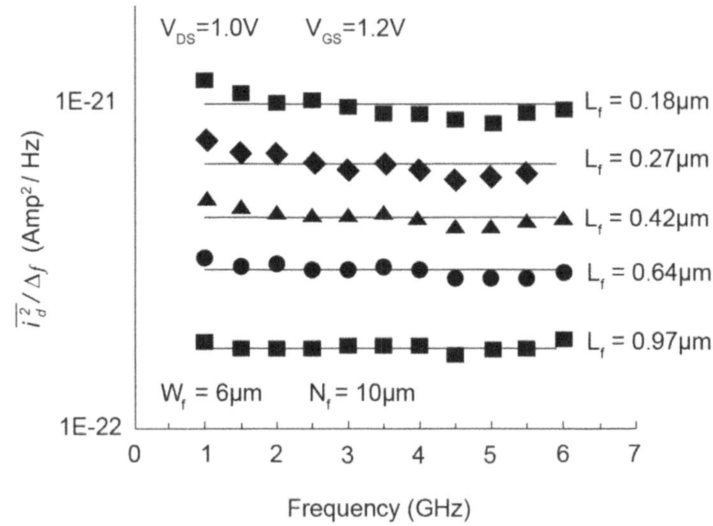

Figure 6.11. *Evolution of $\overline{|\iota_d|^2}/df$ as a function of frequency for different gate lengths according to (Chen et al. 2001)*

7.5. Using the relationships found in 7.2, we can write:

$$\overline{\iota_1 \cdot \iota_2^*} = \overline{(-Y_{11} \cdot e + \iota) \cdot (-Y_{21}^* \cdot e^*)} = -Y_{21}^* \cdot \overline{(-Y_{11} \cdot e + Y_{Cor} \cdot e + \iota_n) \cdot e^*}$$
$$= -Y_{21}^* \cdot (-Y_{11} + Y_{Cor}) \cdot 4kT \cdot df \cdot R_N$$

For this expression to be compatible with Van der Ziel's model, when $Y_{21} = g_m$, that is to say at low frequency, it is necessary that the term $-Y_{11} + Y_{Cor}$ be an imaginary number. For this, it is necessary that $G_{Cor} =$

$Re(Y_{11}) = \frac{R_i \cdot C_{gs}^2 \cdot \omega^2}{1+R_i^2 \cdot C_{gs}^2 \cdot \omega^2}$. At the same time, we know that $G_{Cor} = \frac{F_{eMin}}{2 \cdot R_N} - G_{Sopt}$. We can therefore use the equality $\frac{F_{eMin}}{R_N} - G_{Sopt} = \frac{R_i \cdot C_{gs}^2 \cdot \omega^2}{1+R_i^2 \cdot C_{gs}^2 \cdot \omega^2}$ to check the consistency of measurements in noise or to refine the value of R_i, which is generally difficult to determine. It should be noted that for the MOSFETs some authors use the notation R_{gs} to designate R_i.

For the proposed MOSFET at 2 GHz, we have: $NF_{Min} = 0.37\,dB$, $R_N = 60\Omega$, $\Gamma_{Sopt} = 0.93/5°$, $g_m = 57mS$, $C_{gs} = 125fF$, $C_{gd} = 41fF$, $R_i = 2\Omega$.

From $\Gamma_{Sopt} = 0.93/5°$, we obtain: $y_{Sopt} = 0.0363 - 0.0436j$ and $g_{Cor} = 0.000246$, i.e. after denormalization $G_{Cor} = 4.9\mu S$.

At the same time, the real part of Y_{11} is 4.93 µS. We can therefore consider that we are in the case where $-Y_{11} + Y_{Cor}$ is an imaginary number and therefore that the Van der Ziel model is applicable to this MOSFET.

Consequently, the term $\overline{\iota_1 \cdot \iota_2^*}$ is equal to $j \cdot g_m \cdot (B_{11} + B_{Sopt}) \cdot 4kT \cdot df \cdot R_N$. The term B_{11}, imaginary part of Y_{11} is of the form $j \cdot (C_{gs} + C_{dg}) \cdot \omega$ provided that $1 \gg R_i^2 \cdot C_{gs}^2 \cdot \omega^2$.

With regard to the term B_{Sopt}, to know if it evolves linearly with the pulsation, we can notice that:

$$Arg(\Gamma_{Sopt}) = -Arctg\left(\frac{2 \cdot b_{Sopt}}{1 - g_{Sopt}{}^2 - b_{Sopt}{}^2}\right)$$

If $1 > g_{Sopt}$et $1 > b_{Sopt}$, which is the case at low frequencies, then $Arg(\Gamma_{Sopt}) = -2 \cdot b_{Sopt}$. This means that if the phase Γ_{Sopt} varies linearly with the frequency, then B_{Sopt} varies linearly with the frequency.

In (Asgaramet *et al.* (2004) and Gao (2004), we find the curves presenting $Arg(\Gamma_{Sopt})$ as a function of the frequency reproduced in Figures 6.12 and 6.13, where we clearly see in both cases that the phase of Γ_{Sopt} varies linearly with the frequency.

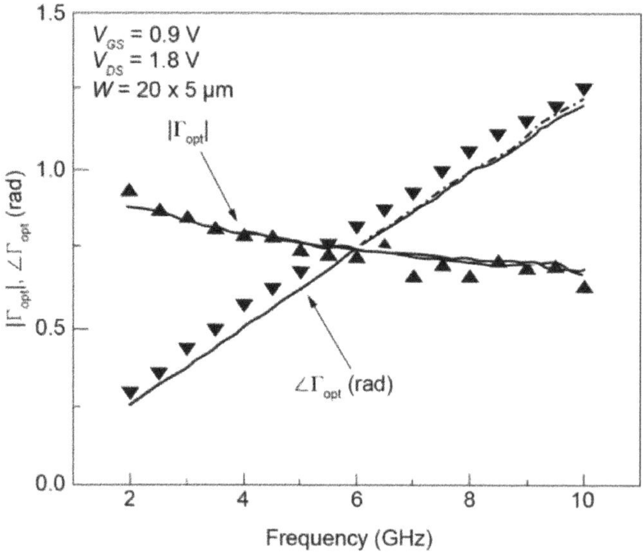

Figure 6.12. *$Arg(\Gamma_{Sopt})$ as a function of the frequency (Asgaram et al. 2004)*

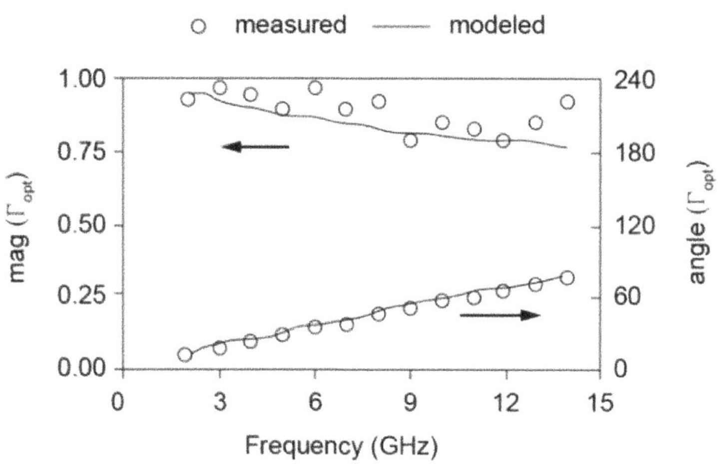

Figure 6.13. *$Arg(\Gamma_{Sopt})$ as a function of the frequency (Gao 2004)*

Thus, we are in the case where B_{Sopt} varies linearly with the frequency and this means that the Van der Ziel model $\overline{i_g^* \cdot i_d} = j \cdot 4kT \cdot df \cdot \omega \cdot C_0$ is applicable for these MOSFETs up to at least 10 GHz.

In this case, the term $(B_{11} + B_{Sopt})$ is about 2 mS at 2GHz with $B_{11} = 2{,}09\ mS\ and\ B_{Cor} = 87\ \mu S$.

7.6. Using the results of 7.2, we obtain:

$$\overline{|\imath_1|^2} = \overline{|-Y_{11}e + \imath|^2} = \overline{|(-Y_{11} + Y_{Cor}) \cdot e + \imath_n|^2}$$
$$= 4kT \cdot df \cdot (|-Y_{11} + Y_{Cor}|^2 \cdot R_N + G_n)$$

Using the answer to question 7.5, we can rewrite $\overline{|\imath_1|^2}$ as:

$$\overline{|\imath_1|^2} = 4kT \cdot df \cdot (B_{11} + B_{Sopt})^2 \cdot R_N + 4kT \cdot df \cdot G_n$$

We have already seen that $B_{11} + B_{Sopt}$ varies in ω so $\overline{|\imath_1|^2}$ varies in ω^2 and is therefore in agreement with Van der Ziel's model which states that $\overline{|\imath_{gtot}|}^2 = \overline{|\imath_{g0}|}^2 + \overline{|\imath_g|}^2 = \overline{|\imath_{g0}|}^2 + 4kT \cdot df \cdot \frac{(\omega \cdot C_0)^2}{g_m}$.

For a MOSFET at low frequencies, we can use the Van der Ziel model and write:

$$\overline{|\imath_2|^2} = 4kT \cdot df \cdot {g_m}^2 \cdot R_N$$
$$\overline{|\imath_1|^2} = 4kT.df \cdot \left(B_{11} + B_{Sopt}\right)^2 \cdot R_N + 4kT \cdot df \cdot G_n$$
$$= 4kT \cdot df \cdot \left[(C_{gg} - C_{Sopt})^2\omega^2 \cdot R_N + G_n\right]$$
$$\overline{\imath_1 \cdot \imath_2^*} = j \cdot g_m \cdot (B_{11} + B_{Sopt}) \cdot 4kT \cdot df \cdot R_N$$
$$= j \cdot g_m \cdot (C_{gg} - C_{Sopt}) \cdot \omega \cdot 4kT \cdot df \cdot R_N$$

with:

$$C_{gg} = C_{gs} + C_{gd} \text{ and } C_{Sopt} = -\frac{B_{Sopt}}{\omega}$$

However, we must ensure that $Re(Y_{11}) = \frac{R_i \cdot C_{gs}^2 \cdot \omega^2}{1 + R_i^2 \cdot C_{gs}^2 \cdot \omega^2} = G_{Cor} = \frac{F_{eMin}}{2 \cdot R_N} - G_{Sopt}$.

With regard to the part $\overline{|\imath_{g0}|}^2 = 4kT \cdot df \cdot G_n$, we can express G_n in the form:

$$G_n = R_N \cdot \left(G_{Sopt}^2 - G_{Cor}^2\right) = \frac{F_{eMin}}{2} \cdot \left(G_{Sopt} - G_{Cor}\right)$$

$$= \frac{F_{eMin}}{2} \cdot \left(G_{Sopt} - \left(\frac{F_{eMin}}{2 \cdot R_N} - G_{Sopt}\right)\right)$$

$$G_n = F_{eMin} \cdot \left[G_{Sopt} - \frac{F_{eMin}}{4 \cdot R_N}\right]$$

For our MOSFET at 2 GHz, this leads to $G_n = 32\mu S$ while $(B_{11} + B_{Sopt})^2 \cdot R_N = 240\ \mu S$. So we see that at 2 GHz, the contribution of G_n to $\overline{|\iota_1|^2}$ is low. Assuming that G_n is independent of frequency (that is, due to white noise), the contributions of Gn and $(B_{11} + B_{Sopt})^2 \cdot R_N$ to $\overline{|\iota_1|^2}$ are identical at 730 MHz (frequency for which the excess noise, in $^1/_f$ or in $^1/_{f^2}$, can be neglected in silicon) and the contribution of G_n to $\overline{|\iota_1|^2}$ will be only 1% at 7.3 GHz.

Conclusion

Electronic Noise and its Measurement

In this book, the formulas necessary to understand the noise behavior of two-ports at radio frequencies have been demonstrated. These rigorous demonstrations make it possible to fully understand the meaning of the various terms appearing in a formula, at the cost of a few manipulations on the analytical expressions used. Nevertheless, the mathematical background required is very reasonable in terms of the lessons learned and, in any case, basic for a young researcher or a recent graduate.

These demonstrations have been conducted for the most part using the admittance matrix, the noise representation using one noise current source and one noise voltage source at the input of the two-port and the classical rules of analysis of the linear electronic circuits which also apply in the context of a noise analysis. The only precaution to be taken appears when it is necessary to move on to noise powers; it is a question of taking into account the correlation between sources which can appear because of the topology of the studied circuit. This general principle is applicable to any circuit, but can quickly become complicated in the writing and handling of different quantities when the circuit has any topology. In this case, it is better to use a complete matrix formalism like the one proposed by (Dobrowolski 1989).

Chapters 4 and 5 present in detail the methodology and the various steps to be carried out to measure a factor on 50 Ω or to characterize in noise any device with two accesses. These two opportunities allow us to see the crucial role of the Friis formula and the need to take into account the available power gain of the object under test. In these chapters, only the problem of measurement uncertainties has not been addressed but the interested reader can refer to the work of (Rohde and Schwarz 2018) or (Keysight 2017b) to evaluate this.

In Chapter 1, the physical causes of background noise were presented but a presentation of their consequences on the behavior of linear transistors was not made; we went directly to an equivalent representation of the noise with two correlated generators seen on the transistor's accesses. The reader wishing to have a synthetic view of the elementary noise sources that appear in a bipolar or field effect transistor can refer to Chapter 7 of (Vendelin *et al.* 2005).

The exercises proposed in Chapter 6, very different from what is found in the literature on noise, make it possible to use the knowledge acquired by reading this book and to establish new interesting results to look deeper into noise phenomena in radio frequency circuits.

Appendix 1

Admittance Parameters of a Two-Port

A1.1. Introduction

Like any matrix description, the admittance parameters are only defined for a linear device such as a passive device or a semiconductor component linearized around an operating point. As this appendix is intended to complete this book on noise, we will only consider the case of two-ports.

A1.2. Definition of admittance parameters

Consider the two-port of Figure A1.1, the admittance parameters aim at establishing the linear relationships that exist between the excitations in voltages V_1 and V_2 and currents I_1 and I_2 which result from them. Using a matrix notation, we can write equations [A1.1]:

$$\begin{vmatrix} I_1 \\ I_2 \end{vmatrix} = \begin{bmatrix} Y_{11} & Y_{12} \\ Y_{21} & Y_{22} \end{bmatrix} \cdot \begin{vmatrix} V_1 \\ V_2 \end{vmatrix} \text{ or } [I] = [Y] \cdot [V] \qquad [A1.1]$$

The terms Y_{ij} are the admittance parameters of the two-port. Each of these terms can be expressed in the form given by [A1.2]:

$$Y_{ij} = \left. \frac{I_i}{V_j} \right|_{V_i=0} \qquad [A1.2]$$

The terms Y_{11} and Y_{22} are "true" admittances, while Y_{21} and Y_{12} are transadmittances or voltage-related current generators that report a transfer, respectively, from port 1 to port 2 and port 2 to port 1.

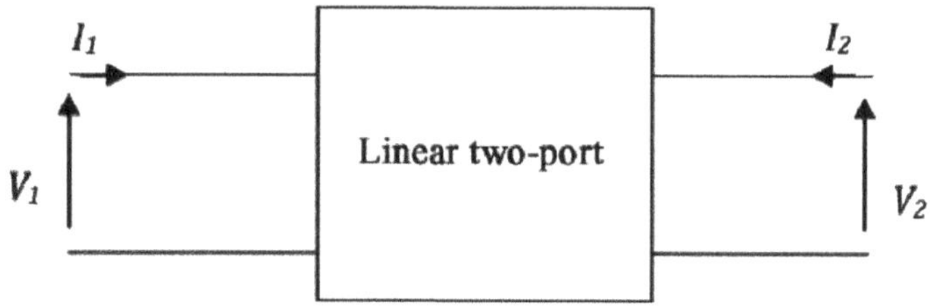

Figure A1.1. *Representation of a linear two-port*

A1.3. Interest of admittance parameters

Like any matrix formulation, the admittance matrix has the advantage that, when we know the internal structure of the two-port or its equivalent circuit, the calculation of Y_{ij} is simpler than facing the problem as a whole. Indeed, to express, for example for Y_{11} and Y_{21}, we must make $V_2 = 0$, simplifying in general the equivalent circuit of the two-port and the computations.

Secondly, if we use the flow graph and Mason's rule, we can write any equation involving this two-port without having to make calculations resulting from the writing of linear equations and substitutions which can to be the cause of many mistakes. This point is addressed in Appendix 3.

A1.4. Equivalent diagram of a two-port known by its admittance matrix

The equivalent "natural" diagram of a two-port with a known admittance matrix is given in Figure A1.2. There are other possible representations but this, although not always adapted to the internal topology of the two-port, is the simplest and is easy to handle.

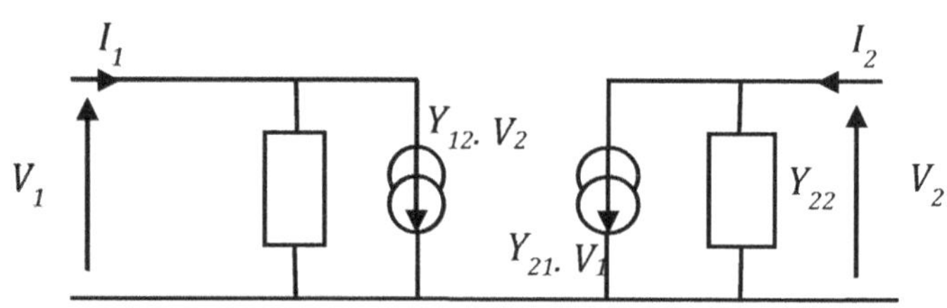

Figure A1.2. *Equivalent diagram of a two-port of known admittance matrix [Y]*

A1.5. Flow graph of a two-port known by its admittance matrix

The flow graph is a graphical representation of equations [A1.1]. It contains the following elements:

– nodes that are the excitations and effects produced, in our case, the voltages V_1 and V_2 and currents I_1 and I_2 produced by these voltages;

– oriented transfers – from the cause towards the effect – which translate the influences between nodes in the device.

The flow graph of a two-port known by its admittance matrix is represented on Figure A1.3, where we find the two equations of [A1.1] given by [A1.3] after development.

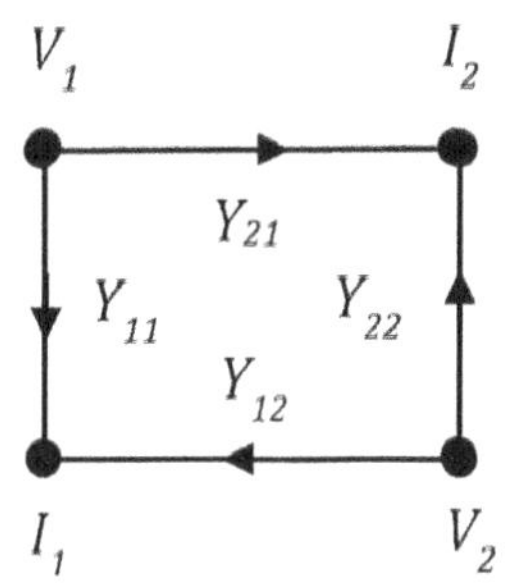

Figure A1.3. *Flow graph of a two-port of known admittance matrix [Y]*

$$I_1 = Y_{11} \cdot V_1 + Y_{12} \cdot V_2$$

$$I_2 = Y_{21} \cdot V_1 + Y_{22} \cdot V_2 \qquad [A1.3]$$

A1.6. Generator + two-port + charge

Quite classically in electronics, a two-port which is used to carry out a processing on a signal is associated with a generator which delivers, at the input of the two-port, the signal to be processed and a charge, placed at the output of the two-port, which retrieves the signal after treatment. A flow graph can be used to represent this association which is described by equations [A1.4] which describe the constitution of the elements used, and by equations [A1.5] which describe the connections made between the elements used. This corresponding flow graph is given in Figure A1.

$$I_1 = Y_{11} \cdot V_1 + Y_{12} \cdot V_2 \qquad [A1.4a]$$

$$I_2 = Y_{21} \cdot V_1 + Y_{22} \cdot V_2 \qquad [A1.4b]$$

$$V_S = E - Z_S \cdot I_S \qquad [A1.4c]$$

$$V_L = -Z_L \cdot I_L \qquad [A1.4d]$$

The generator contains an ideal voltage source E in series with an internal impedance Z_S.

The charge has a value Z_L.

$$I_1 = I_S \qquad [A1.5a]$$

$$V_1 = V_S \qquad [A1.5b]$$

$$I_L = I_2 \qquad [A1.5c]$$

$$V_2 = V_L \qquad [A1.5d]$$

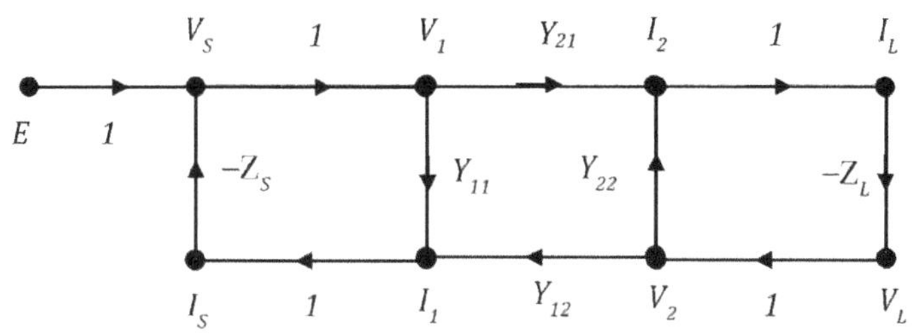

Figure A1.4. *Flow graph of generator, two-port and charge association*

In the graph of Figure A1.4, note that node E is the causal node; that is to say, the one that makes all other magnitudes non-zero. From this graph, we can express all the quantities that characterize this association, as we will see in Appendix A3.

A1.7. Conclusion on the admittance matrix

The admittance matrix, despite the interest it presents when it comes to conducting literal calculations, is given little appreciation in electronics because it is difficult to implement from the point of view of measurements. Indeed, it requires placing a short circuit at the output ($V_2 = 0$), to set an input excitation V_1 and to measure two currents I_1andI_2 to then repeat the same thing, but with $V_1 = 0$. These short circuits are worrying and the measurement of current is difficult to implement.

Appendix 2

S Parameters of a Two-Port

A2.1. Introduction

Like any matrix description, the *S* parameters are defined only for a linear device such as a passive device or a semiconductor component linearized around an operating point. As this appendix is intended to complement this noise study, we will only consider the case of two-ports and normalization on ports with a real reference impedance $R_0 = 50\ \Omega$.

A2.2. Definition of *S* parameters

Consider the two-port of Figure A2.1, the *S* parameters aim to establish the linear relations that exist between the excitations, called the incident power waves a_1 and a_2, and the reflected power waves b_1 and b_2 that result (Kurokawa 1965, Anderson 1967). Using a matrix notation, we can write equations [A2.1]:

$$\begin{vmatrix} b_1 \\ b_2 \end{vmatrix} = \begin{bmatrix} S_{11} & S_{12} \\ S_{21} & S_{22} \end{bmatrix} \cdot \begin{vmatrix} a_1 \\ a_2 \end{vmatrix} \text{ or } [B] = [S].[A] \qquad \text{[A2.1]}$$

The terms S_{ij} are the *S* parameters of the two-port. Each of these terms can be expressed in the form given by equation [A2.2]:

$$S_{ij} = \left.\frac{b_i}{a_j}\right|_{a_i=0} \qquad \text{[A2.2]}$$

The terms S_{11} and S_{22} are called reflection coefficients, while S_{21} and S_{12} are transmission coefficients.

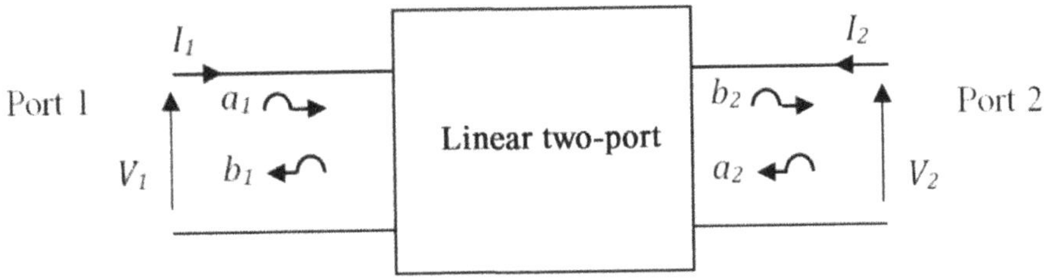

Figure A2.1. *Representation of the power waves of a linear two-port*

A2.3. Passage equations of power waves and current / voltage

To understand what these power waves are and therefore the S parameters, the simplest way is to start from the power exchange between a generator E, Z_S and a load Z_L as shown in Figure A2.2.

The power P_L collected in the load is given by equation [A2.3]:

$$P_L = Re(V_L \cdot I_L^*) = Re\left(\left[\frac{E.Z_L}{Z_L+Z_G}\right] \cdot \left[\frac{E^*}{(Z_L+Z_G)^*}\right]\right) = |E|^2 \cdot \frac{Re(Z_L)}{|Z_L+Z_G|^2} \quad [A2.3]$$

The power wave formalism (Kurokawa 1965) proposes to define for the load two power waves and a coefficient of proportionality between these two waves:

– an incident power wave for the load a_L such that the incident power seen by the load is $|a_L|^2$;

– a power wave reflected by the load b_L that the power reflected by the load is $|b_L|^2$;

– a reflection coefficient of the load Γ_L with respect to the reference impedance $R_0 = 50\ \Omega$ whose expression is given by equation [A2.4]:

$$\Gamma_L = \frac{Z_L - R_0}{Z_L + R_0} = \frac{b_L}{a_L} \quad [A2.4]$$

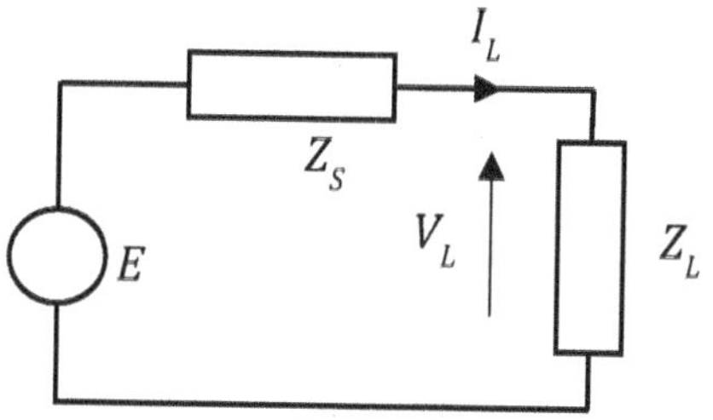

Figure A2.2. *Generator connection*

Under these conditions, the power P_L collected in the load is given by equation [A2.5]:

$$P_L = |a_L|^2 - |b_L|^2 = |a_L|^2 \cdot (1 - |\Gamma_L|^2) \quad \text{[A2.5]}$$

If a reference generator E, R_0 is used to define these power waves, equations [A2.3] and [A2.5] are rewritten, respectively, in the form of equations [A2.6] and [A2.7]:

$$P_L = Re(V_L \cdot I_L^*) = |E|^2 \cdot \frac{Re(Z_L)}{|Z_L+R_0|^2} \quad \text{[A2.6]}$$

$$P_L = |a_L|^2 \cdot (\frac{4 \cdot R_0 \cdot Re(Z_L)}{|Z_L+R_0|^2}) \quad \text{[A2.7]}$$

By identification, we find the expression of a_L given by equation [A2.8]:

$$a_L = \frac{E}{2.\sqrt{R_0}} = \frac{V_L+R_0.I_L}{2.\sqrt{R_0}} \quad \text{[A2.8]}$$

Using [A2.4] and [A2.8], we can express b_L as in equation [A2.9]:

$$b_L = \frac{V_L-R_0 \cdot I_L}{2 \cdot \sqrt{R_0}} \quad \text{[A2.9]}$$

By using the magnitudes normalized to the reference impedance $R_0 = 50\ \Omega$ as explained in equation [A2.13], we can rewrite equations [A2.4], [A2.8] and [A2.9] in the forms given by equations [A2.10] to [A2.12]:

$$\Gamma_L = \frac{z_L-1}{z_L+1} = \frac{b_L}{a_L} \quad \text{[A2.10]}$$

$$a_L = \frac{v_L + i_L}{2} \quad [A2.11]$$

$$b_L = \frac{v_L - i_L}{2} \quad [A2.12]$$

$$v_L = \frac{V_L}{\sqrt{R_0}}, \; i_L = I_L \cdot \sqrt{R_0}, \; z_L = \frac{Z_L}{R_0} \quad [A2.13]$$

A reflection coefficient load Γ_L can be described by the flow graph given in Figure A2.3.

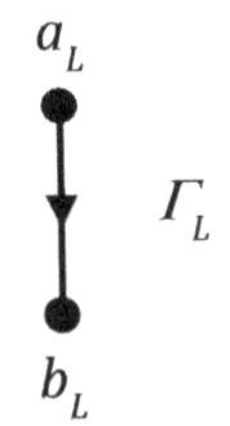

Figure A2.3. *Flowgraph of a reflection coefficient load Γ_L*

The relations equations [A2.11] and [A2.12] constitute, associated with the normalization equations [A2.13], the relations of passage between the power waves and the normalized currents / voltages.

The same power wave formalism can be applied to the generator by defining:

– an incident power wave for the generator a_S;

– a power wave reflected by the generator b_S;

– a reflection coefficient of the generator $\Gamma_S = \frac{Z_S - R_0}{Z_S + R_0}$ with respect to the reference impedance $R_0 = 50\ \Omega$.

The only difference with a load is that the generator will provide a reflected wave b_S even in the absence of incident wave a_S because it is a generator. So we have $b_S = b_{Sint} + \Gamma_S \cdot a_S$. This internally generated power wave b_{Sint} can be calculated by charging this generator with the reference impedance $R_0 = 50\ \Omega$. Taking again the notations of Figure A2.2 for $Z_L = R_0$, one obtains, by using equations [A2.3] and [A2.5] with $\Gamma_L = 0$, the equality given by equation [A2.14]. This is due to the fact that all the power waves are defined with respect to the same reference $R_0 = 50$ and that the incident wave for the charge a_L is the wave reflected by the generator b_S – or which emerges from the generator if we prefer to say so – which is itself equal to b_{Sint} because $b_L = a_S = 0$ because of $\ _L = 0$:

$$P_L = |E|^2 \cdot \frac{Re(R_0)}{|R_0+Z_G|^2} = |a_L|^2 = |b_S|^2 = |b_{Sint}|^2 \qquad \text{[A2.14]}$$

One can write equation [A2.15] which gives the relation between b_{Sint} and $e = \frac{E}{\sqrt{R_0}}$ normalized value of E:

$$b_{Sint} = \frac{e}{1+z_S} \qquad \text{[A2.15]}$$

The flow graph of a generator is given in Figure A2.4.

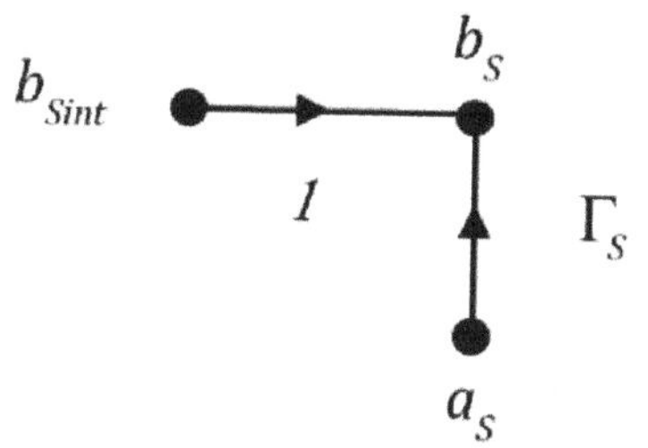

Figure A2.4. *Flowgraph of a generator* E, Γ_S

A2.4. Physical meaning of *S* parameters

To understand the physical meaning of the S_{ij}, it is enough to generalize the passage equations [A2.11] and [A2.12] for a two-port for these ports 1 and 2 by using the expressions [A2.16]:

$$a_i = \frac{v_i+i_i}{2} \qquad \text{[A2.16a]}$$

$$b_i = \frac{v_i-i_i}{2} \qquad \text{[A2.16b]}$$

$$v_i = \frac{V_i}{\sqrt{R_0}}, \; i_i = I_i.\sqrt{R_0}, \qquad \text{[A2.16c]}$$

Under these conditions, the condition $a_2 = 0$, which makes it possible to say that $S_{11} = \frac{b_1}{a_1}\Big|_{a_2=0}$ and $S_{21} = \frac{b_2}{a_1}\Big|_{a_2=0}$, takes on its full meaning because it imposes $v_2 + i_2 = 0$ that we can write $V_2 + R_0.I_2 = 0$. That means, considering the current I_2 returning that enters in port 2 as shown in Figure A2.1, that the output is loaded by the reference impedance R_0.

The two-port behaves therefore, seen from port 1, as a dipole because the port 2 is closed on R_0. The term S_{11} is therefore the reflection coefficient of the port 1 when the port 2 is loaded by the reference impedance R_0. As such, it can be written in the form given by equation [A2.17] by saying that $Z_{IN1}|_{R_0}$ is the impedance seen on port 1 when port 2 is loaded by the reference impedance R_0:

$$S_{11} = \frac{Z_{IN1}|_{R_0} - R_0}{Z_{IN1}|_{R_0} + R_0} \qquad [A2.17]$$

Thanks to the symmetry of the ports 1 and 2, we can immediately give the meaning of S_{22}: S_{22} as is the reflection coefficient of port 2 when port 1 is loaded by the reference impedance R_0 and it can be written in the form given by [A2.18], saying that $Z_{IN2}|_{R_0}$ is the impedance seen on port 2 when port 1 is loaded by the reference impedance R_0

$$S_{22} = \frac{Z_{IN2}|_{R_0} - R_0}{Z_{IN2}|_{R_0} + R_0} \qquad [A2.18]$$

For the term $S_{21} = \frac{b_2}{a_1}\Big|_{a_2=0}$, the condition of port 2 loaded by R_0 is still necessary, but we can also notice that $b_2 = v_2$. For port 1, it is necessary to create an incident wave $a_1$and there is no condition to respect; we can then choose a generator E, R_0. Under these conditions, the wave b_S that emerges from the generator is only due to b_{Sint} because $\Gamma_S = 0$. This wave b_S is also the incident wave a_1 for port 1 of the two-port and using [A2.15], we can write $a_1 = \frac{e}{2}$.

Under these conditions, the expression of S_{21} is given by [A2.19]. This means that S_{21} is equal to twice the composite voltage gain of the two-port, in the direction port 1 to port 2, calculated for an internal impedance of the generator and a load impedance both equal to the reference impedance R_0:

$$S_{21} = \left.\frac{2.v_2}{e}\right|_{\Gamma_L=\Gamma_S=0} = \left.\frac{2.V_2}{E}\right|_{\Gamma_L=\Gamma_S=0} = 2.\,G_{VC,1>2}\big|_{Z_L=Z_S=50\Omega} \qquad \text{[A2.19]}$$

In the same way, S_{12} is equal to twice the composite voltage gain of the two-port, in the direction port 2 to port 1, calculated for an internal impedance of the generator and a load impedance both equal to the reference impedance R_0. Its expression is given by equation [A2.20]:

$$S_{12} = \left.\frac{2.v_1}{e}\right|_{\Gamma_L=\Gamma_S=0} = \left.\frac{2.V_1}{E}\right|_{\Gamma_L=\Gamma_S=0} = 2.\,G_{VC,2>1}\big|_{Z_L=Z_S=50\Omega} \qquad \text{[A2.20]}$$

These relations [A2.17] to [A2.20] are very practical when one wants to calculate, from the internal constitution of a two-port, its S parameters but they are, strangely, rarely used.

A2.5. Flow graph of a two-port known by its *S* matrix

The flow graph is a graphical representation of equations [A2.1]. It contains the following elements:

– nodes that are the excitations and effects produced; in our case, the incident power waves a_1 and a_2 and the reflected power waves b_1 and b_2;

– oriented transfers – from the cause towards the effect – which translate the influences between nodes in the device and which are here, S_{ij}.

The flow graph of a two-port known by its S matrix is represented in Figure A2.5, where we find the two equations of [A2.1] given by equations [A2.21] after development.

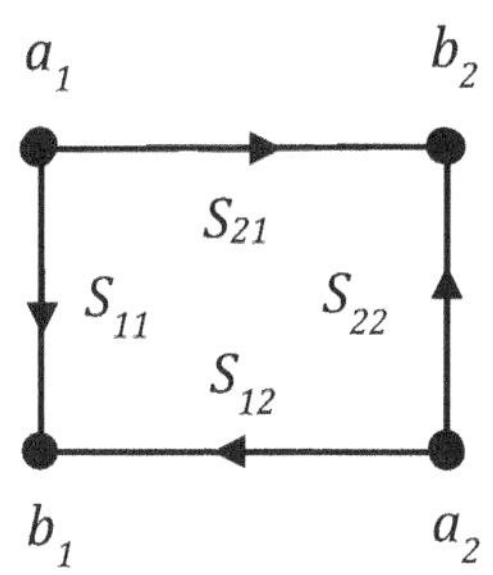

Figure A2.5. *Flow graph of a two-port known by its S matrix*

$$b_1 = S_{11}.a_1 + S_{12}.a_2 \qquad [A2.21a]$$

$$b_2 = S_{21}.a_1 + S_{22}.a_2 \qquad [A2.21b]$$

A2.6. Generator + two-port + load

As for the matrix admittance, this association which is described by the relations [A2.22] which describe the constitution of the elements used – with two equations for the two-port, one for the generator and one for the load – and by the four relations [A2.23] which describe the connections made between the elements used. This corresponding flow graph is given in Figure A2.6:

$$b_1 = S_{11} \cdot a_1 + S_{12} \cdot a_2 \qquad [A2.22a]$$

$$b_2 = S_{21} \cdot a_1 + S_{22} \cdot a_2 \qquad [A2.22b]$$

$$b_S = b_{Sint} + \Gamma_S \cdot a_S \qquad [A2.22c]$$

$$b_L = \Gamma_L \cdot a_L \qquad [A2.22d]$$

$$a_1 = b_S \qquad [A2.23a]$$

$$b_1 = a_S \qquad [A2.23b]$$

$$a_L = b_2 \qquad [A2.23c]$$

$$a_2 = b_L \qquad [A2.23d]$$

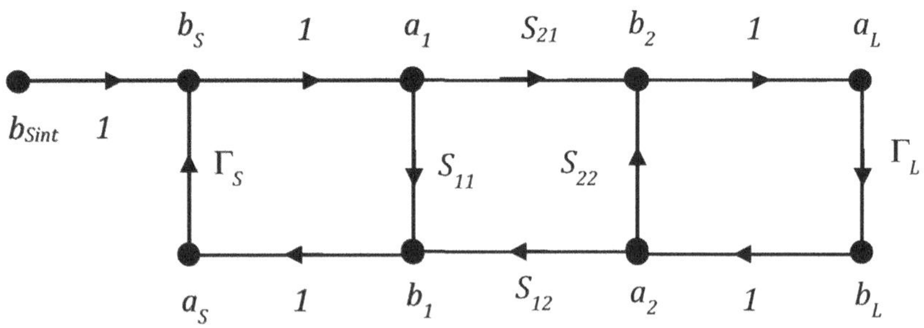

Figure A2.6. *Flow graph of generator, two-port and load association*

In the graph of Figure A2.6, note that only the node b_{Sint} is the causal node. From this graph, we can express all the quantities that characterize this association, as we will see in Appendix 3.

It can be seen that the shape of the graph in Figure A2.6 is exactly the same as in Figure A1.4 in Appendix 1. This is normal because it is the same circuit simply described with two different languages, but the connections, made between the generator, the two-port and the load, remain basically the same. Of course, the same form is obtained if the impedance matrix is used.

A2.7. Conclusion on the *S* matrix

The *S* matrix is appreciated in electronics and microwaves because it lends itself well to measurement. It is a question of measuring voltages on a two-port by exciting it on a port with a generator $R_0 = 50\ \Omega$ and by placing on the other port a load $R_0 = 50\ \Omega$. On the other hand, the S-matrix is not very well adapted to the association of objects, even if solutions exist (Dobrowolski 1989).

Since the *S* matrix and the Y matrix are matrices in which all the coefficients have the same unit, there are matrix relationships between these two matrices. These equations are given by [A2.24] and [A2.25]:

$$[S] = [Id + Y_N]^{-1} \cdot [Id - Y_N] \qquad \text{[A2.24]}$$

$$[Y_N] = [Id + S]^{-1} \cdot [Id - S] \qquad \text{[A2.25]}$$

where:

– $[Id]$ is the identity matrix of the same rank as [S];

– $[Y_N] = R_0 . [Y]$ is the admittance matrix normalized to R_0.

These equations can be shown using a matrix formulation of passage equations [A2.16].

Appendix 3

Flow Graph and Mason's Rule

A3.1. Introduction

Mason's rule (Mason 1956) is a method for solving a system of linear equations based on the observation of the flow graph representing the linear equations of the system. These equations are of two types:

– those which describe the constitution of the elements of the system;

– those that describe how these elements relate to each other.

This method, highly appreciated in automatics, can be used in electronics and has the advantage of minimizing calculations to find the relationship between two electrical quantities in a circuit.

In the remainder of this Appendix, we will state Mason's rule in all its generality and then apply it only to two-ports that can be connected to a load and / or a generator. This will allow us to show how certain expressions of quantities used in this work can be established.

A3.2. Statement of Mason's Rule

Recall that a flow graph contains nodes N_i which represent the quantities used in the system of linear equations and t_{ij} oriented transfers from the cause to the effect which indicate the influence between the node j and the node i. These influences may be related to the constitution of an element, an electronic component for us, or describe a connection made in the system, a circuit combining several components in our case.

In a flow graph, some nodes have a special role because they correspond to the excitations of the system which, for us, are free sources. This is called a causal node.

Mason's rule says that, in such a graph, the transfer between a causal node N_{iC} of the system and any node of the graph N_j is written in the form given by equation [A3.1]:

$$T_{j,iC} = \frac{N_j}{N_{iC}} = \frac{\sum_{k=1}^{k} T_{k,j,iC} \cdot \Delta_k}{\Delta_{Mason}} \quad \text{[A3.1]}$$

The terms that appear in equation [A3.1] have the following definition:

The index k corresponds to the number of direct paths to go from the departure node N_{iC} to the arrival node N_j. Path is understood to mean the possibility of going from N_{iC} to N_j by using transfers t_{lm} while respecting the direction of the transfer. By direct, we mean that during the course of the path k, it is forbidden to pass more than once by the same node;

– the transfer $T_{k,j,iC}$ is the transfer associated with the path k; i.e. the product of the transfers t_{lm} to go from N_{iC} to N_j;

– the term Δ_{Mason} is the "delta of Mason" whose calculation will be explained later in this section;

– the term Δ_k is "Mason's delta" relative to the path k and it is calculated by eliminating from it all the combinations showing t_{lm} transfers via the path k.

The term Δ_{Mason} is calculated, as indicated in formula [A3.2]:

$$\Delta_{Mason} = 1 - \sum_{on\ all\ nodes}(Node\ transfer) + \sum_{on\ all\ nodess\ 2\ to\ 2\ disjoined}(\prod_{22}(Node\ transfer) - \sum_{on\ all\ nodes\ 3\ to\ 3\ disjoined}(\prod_{33}(Node\ transfer) + \sum_{on\ all\ nodes\ 4\ to\ 4\ disjoined}(\prod_{44}(Node\ transfer) - \ldots \quad \text{[A3.2]}$$

To explain the meaning of expression [A3.2], we must explain the notion of loop. A loop, in a flow graph, is a particular path in that, starting from any node of this loop, we can return to it by using transfers from the graph respecting the direction of the transfer. The loop transfer is the product of t_{lm} transfers followed during the course of the loop. Loops are called disjointed if they do not have transfers in common. To avoid making mistakes, it is strongly recommended to make no simplification in the graph before applying Mason's rule.

A3.3. Application of Mason's rule to a two-port

A3.3.1. *Introduction*

We will apply this calculation method to express, for a two-port known by its *S* matrix or *Y* matrix, the following quantities:

– the reflection coefficient or admittance, input or output, when the other port is loaded;

– the voltage gain for different input and output load conditions;

– the different power gains.

These quantities were used in this book without sometimes showing where their expression came from.

A3.3.2. *Reflection and admittance coefficient*

To express this type of expression which concerns only one port, since the other one is closed by a load, the flow graph always has the form given in Figure A3.1, in S parameters or in Y parameters.

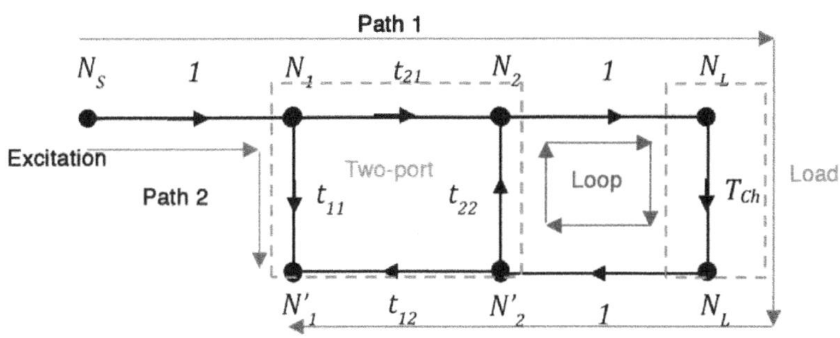

Figure A3.1. *Flow graph of a loaded two-port on a port. For a color version of this figure, see www.iste.co.uk/fouquet/electronics.zip*

On the graph of Figure A3.1, we see:

The two-port represented by the four electrical quantities which characterize it, two for port 1 N_1, N'_1 and two for port 2 N_2, N'_2 as well as the transfers t_{ij} which characterize its operation. These four t_{ij} can be indifferently *S* parameters or *Y* parameters;

A load which will be defined, optionally by $t_{Ch} = \Gamma_L$ ou $t_{Ch} = -Z_L$ according to whether we use *S* parameters or *Y* parameters.

The transfer we are interested in is N_1/N'_1 which concerns only the entry port. The causal node in this graph is N_S which is the one that creates all the other quantities in the graph. We will have to calculate by Mason N_1/N_S and N'_1/N_S.

In Figure A3.1, we see that:

– there is a single loop B_1 in the graph and its transfer is $T_{B1} = 1 \cdot t_{Ch} \cdot 1 \cdot t_{22}$;

– there are two paths in the graph to go from N_S to N'_1 :

- path 1 has at least one common transfer with loop B1 and therefore the transfer is $T_{ch1} = 1 \cdot t_{21} \cdot 1 \cdot t_{Ch} \cdot t_{12}$;

- path 2 has no common transfer with the loop B_1 and therefore the transfer is $T_{ch2} = 1.t_{11}$.

- Δ_{Mason} is therefore $1-(1.t_{Ch}.1.t_{22}.)$, Δ_1 is 1 and Δ_1 is $1 -(1.t_{Ch}.1.t_{22}.)$.

The transfer between N'_1 to N_S is given by [A3.3]:

$$\frac{N'_1}{N_S} = \frac{T_{ch1}.\Delta_1 + T_{ch2}.\Delta_2}{\Delta_{Mason}} = \frac{t_{11}.(1-t_{Ch}.t_{22}) + t_{21}.t_{Ch}.t_{12}.1}{1-t_{Ch}.t_{22}} = t_{11} + \frac{t_{21}.t_{Ch}.t_{12}}{1-t_{Ch}.t_{22}} \quad [A3.3]$$

For transfer N_1/N_S, the same method is applied with a single path to go from N_S to N_1 which does not touch loop B_1; therefore $T_{ch3} = 1$ and $\Delta_3 = 1- t_{Ch}.t_{22}$. The transfer between N_1 to N_S is given by equation [A3.4]:

$$\frac{N_1}{N_S} = \frac{T_{ch3}.\Delta_3}{\Delta_{Mason}} = \frac{1.(1-t_{Ch}.t_{22})}{1-t_{Ch}.t_{22}} = 1 \quad [A3.4]$$

Hence the expression of the sought-after transfer given by equation [A3.5]:

$$\frac{N'_1}{N_1} = t_{11} + \frac{t_{21}.t_{Ch}.t_{12}}{1-t_{Ch}.t_{22}} \qquad [A3.5]$$

So, if we reason in *S* parameters, we can write that:

– the reflection coefficient on port 1, when port 2 is loaded by Γ_L , is:

$$\left.\frac{b_1}{a_1}\right|_{\Gamma_L} = \Gamma_{In} = S_{11} + \frac{S_{21} \cdot S_{12} \cdot \Gamma_L}{1 - \Gamma_L \cdot S_{22}}$$

– the reflection coefficient on port 2, when port 1 is loaded by Γ_S , is:

$$\left.\frac{b_2}{a_2}\right|_{\Gamma_S} = \Gamma_{Out} = S_{11} + \frac{S_{21} \cdot S_{12} \cdot \Gamma_S}{1 - \Gamma_S \cdot S_{11}}$$

So, if we reason in *Y* parameters, we can write that:

– the admittance seen on port 1, when port 2 is loaded by Z_L (Ohm's law on the load is written $V_L = -I_L \cdot Z_L$ with the direction chosen for $I_L = I_2$), equal to:

$$\left.\frac{I_1}{V_1}\right|_{Z_L} = Y_{In} = Y_{11} - \frac{Y_{21} \cdot Y_{12} \cdot Z_L}{1 + Z_L \cdot Y_{22}} = Y_{11} - \frac{Y_{21} \cdot Y_{12}}{Y_L + Y_{22}}$$

– the admittance seen on port 2, when port 1 is loaded by Z_S (Ohm's law on the generator is written $V_S = -I_S \cdot Z_S$ with the direction chosen for $I_S = I_1$), equal to:

$$-\left.\frac{I_2}{V_2}\right|_{Z_S} = Y_{Out} = Y_{22} - \frac{Y_{21} \cdot Y_{12} \cdot Z_S}{1+Z_S \cdot Y_{11}} = Y_{22} - \frac{Y_{21} \cdot Y_{12}}{Y_S+Y_{11}}$$

A3.3.3. *Voltage gains*

To express the different voltage gains, we will only reason with the admittance parameters. In this case, the flow graph to be used is that of Figure A3.2. The transfer to be calculated is $\left.\frac{V_2}{E}\right|_{Z_S,Z_L}$ which is the gain of the two-pole for a condition of excitation and charge which we call the composite voltage gain which thus includes the influence the generator's internal impedance and load.

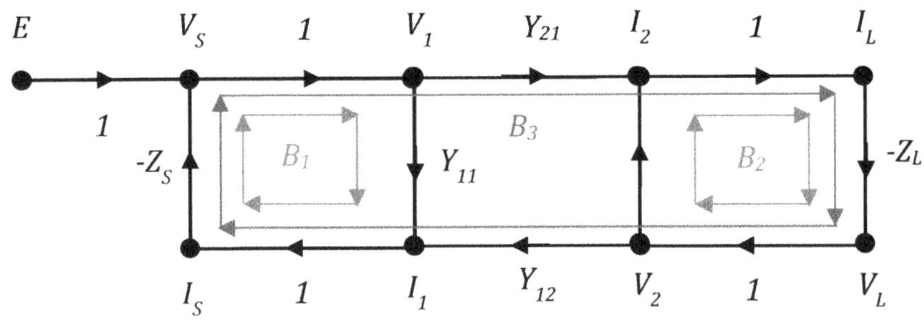

Figure A3.2. *Flow graph of matrix two-port Q admittance [Y] fed by source E, Z_S, and output by Z_S. For a color version of this figure, see www.iste.co.uk/fouquet/electronics.zip*

This flow graph comprises three loops B_1, B_2 and B_3. Only B_1 and B_2 are disjoined therefore the expression of Δ_{Mason} stops at the product of loops 2 to 2 disjoined. Loop transfers are:

– for B_1 : $-Y_{11}{\cdot}Z_S$;

– for B_2 : $-Y_{22}{\cdot}Z_L$;

– for B₃ : $Y_{12}{\cdot}Z_S{\cdot}Y_{21}{\cdot}Z_L$.

Δ_{Mason} relative to this graph is given by equation [A3.6]:

$$\Delta_{Mason} = 1 - \underbrace{(-Y_{11} \cdot Z_S - Y_{22} \cdot Z_L + Y_{12} \cdot Y_{21} \cdot Z_S \cdot Z_L)}_{\text{Sum on all loops of loop transfers}} + \underbrace{((Y_{11} \cdot Z_S) \cdot (Y_{22} Z_L))}_{\text{Sum on all loops 2 to 2 disjoint loop transfers}}$$

$$\Delta_{Mason} = (1 + Y_{11} \cdot Z_S) \cdot (1 + Y_{22} \cdot Z_L) - Y_{12} \cdot Y_{21} \cdot Z_S \cdot Z_L \qquad \text{[A3.6]}$$

To go from E to V_2, there is only one transfer path $T_1 = -Y_{21} \cdot Z_L$. The composite voltage gain of the two-port is given by equation [A3.7]:

$$G_{VC}(Z_S, Z_L) = \left.\frac{V_2}{E}\right|_{Z_S, Z_L} = \frac{-Y_{21} \cdot Z_L \cdot}{(1+Y_{11} . Z_S) \cdot (1+Y_{22} . Z_L) - Y_{12} \cdot Y_{21} \cdot Z_S \cdot Z_L} \qquad \text{[A3.7]}$$

– From this composite voltage gain, we can, if desired, express other voltage gains:

– $G_V(Z_L)$ the own voltage gain of the two-port, obtained by letting $Z_S = 0$ in [A3.7]:

$$G_V(Z_L) = \frac{V_2}{E}\bigg|_{Z_L} = \frac{-Y_{21}.Z_L}{(1+Y_{22}.Z_L)}$$

– G_V the no load own voltage gain, obtained by letting $Z_S = 0$ and $Z_L = +\infty$ in [A3.7]:

$$G_V(Z_L) = \frac{V_2}{E}\bigg|_{Z_L} = \frac{-Y_{21}}{Y_{22}}$$

A3.3.4. *Power gains*

For the expression of the power gains, the work is relevant in S parameters and Y parameters since we used the two forms in this work. We will therefore work with the graph of Figure A3.3 which has the same shape as Figure A3.2; only the notations have been changed to ensure the compatibility of *S* parameters and *Y* parameters. We will express the following gains with this graph:

– the transducer power gain $G_T(Y_S, Y_L) = \frac{P_L}{P_{SAvai}}$ where P_L is the power collected in the load and P_{SAvai} is the power available at the source;

– the available power gain $G_{Avai}(Y_S) = \frac{P_L}{P_{SAvai}}$ where P_L is the power collected in a load admittance which is the conjugated admittance to that seen at the output of the two-port and P_{SAvai} is the power available at the source;

– the power insertion gain $G_i(Y_S, Y_L) = \frac{P_L}{P_L'}$ where P_L is the power collected in the load by inserting the two-port between the source and the load and P_L' is the power collected in the load connected directly to the source.

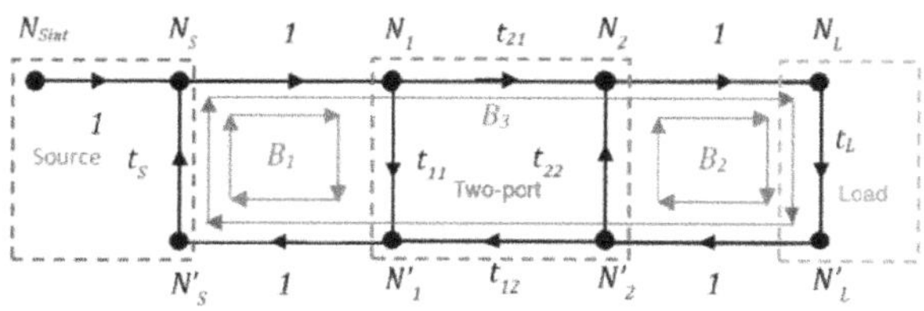

Figure A3.3. *Flow graph of a two-port Q inserted between a source and a load. For a color version of this figure, see www.iste.co.uk/fouquet/electronics.zip*

In Y parameters, the power collected in the load is $P_L = Re(V_L.I_L^*) = |I_L|^2 \cdot Re(1/Y_L)$ with:

$$V_L = N'_L = I_L/Y_L$$

$$I_L = N_L$$

$$N_{Sint} = E$$

$$t_L = -\frac{1}{Y_L}$$

$$t_S = -\frac{1}{Y_S}$$

In S parameters, the power collected at the load is $P_L = |a_L|^2 - |b_L|^2 = |a_L|^2.(1 - |\Gamma_L|^2)$ with:

$$b_L = N'_L = \Gamma_L \cdot a_L$$

$$a_L = N_L$$

$$N_{Sint} = b_{Sint}$$

$$t_L = \Gamma_L$$

$$t_S = \Gamma_S$$

In both cases, we have to calculate the transfer between the causal node N_{Sint} and the arrival node N_L. There is only one direct path between these two nodes and it touches the three loops. We thus have a path transfer which is $T_1 = t_{21}$ and $\Delta_1 = 1$. The transfer sought has the expression given by equation [A3.8]:

$$\frac{N_L}{N_{Sint}} = \frac{t_{21}}{(1-t_{11} \cdot t_S) \cdot (1-t_{22} \cdot t_L) - t_{12} \cdot t_{21} \cdot t_S \cdot t_L} \qquad \text{[A3.8]}$$

For the power collected in the load, we obtain the expression [A3.9] in Y parameters and the expression [A3.10] in S parameters:

$$P_L = \frac{|E|^2 |Y_{21}|^2}{\left|1 + \frac{Y_{11}}{Y_S} + \frac{Y_{22}}{Y_L} - \frac{Y_{12} Y_{21}}{Y_S\, Y_L} + \frac{Y_{11} Y_{22}}{Y_S\, Y_L}\right|^2} \cdot Re(1/Y_L) \qquad \text{[A3.9]}$$

$$P_L = \frac{|b_{Sint}|^2 |S_{21}|^2}{|(1-S_{11}\cdot\Gamma_S)\cdot(1-S_{22}\cdot\Gamma_L)-S_{12}\cdot S_{21}\cdot\Gamma_S\cdot\Gamma_L|^2}\cdot(1-|\Gamma_L|^2) \quad \text{[A3.10]}$$

We now have to calculate P'_L and P_{SAvai}. This is done with the flow graph of Figure A3.4 where the generator is directly connected to a load.

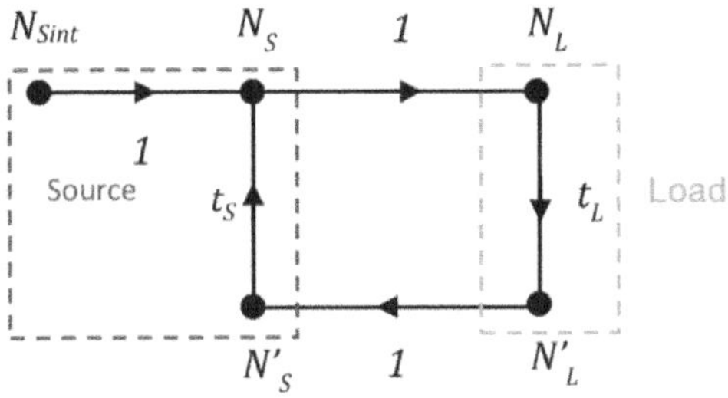

Figure A3.4. *Flow graph of the connection between a source and a load. For a color version of this figure, see www.iste.co.uk/fouquet/electronics.zip*

The flow graph of Figure A3.4 has a single loop B_1, thus the expression of Δ_{Mason} stops at the sum of the loop transfer. The loop transfer is $t_L \cdot t_S$. The Δ_{Mason} relative to this graph is given by equation [A3.11]:

$$\Delta_{Mason} = 1 - t_S \cdot t_L \quad \text{[A3.11]}$$

In both cases, we must calculate the transfer between the causal node N_{Sint} and the arrival node N_L. There is only one direct path between these two nodes and it touches the loop. We thus have a path transfer for which $T_1 = 1$ and $\Delta_1 = 1$.

The transfer sought has the expression given by equation [A3.12]:

$$\frac{N_L}{N_{Sint}} = \frac{1}{1-t_S.t_L} \quad \text{[A3.12]}$$

In Y parameters, the power collected in the load will be $P'_L = Re(V_L \cdot I_L^*) = |V_L|^2 \cdot Re(Y_L)$, with:

$$V_L = N_L = I_L/Y_L$$

$$I_L = N'_L$$

$$N_{Sint} = E$$

$$t_L = Y_L$$

$$t_S = -1/Y_S$$

In S parameters, the power collected in the load is $P'_L = |a_L|^2 - |b_L|^2 = |a_L|^2 \cdot (1 - |\Gamma_L|^2)$, with:

$$b_L = N'_L = \Gamma_L \cdot a_L$$

$$a_L = N_L$$

$$N_{Sint} = b_{Sint}$$

$$t_L = \Gamma_L$$

$$t_S = \Gamma_S$$

For the power collected in the load, we obtain the expression [A3.13] in Y parameters and the expression [A3.14] in S parameters:

$$P'_L = \frac{|E|^2}{\left|1+\frac{Y_L}{Y_S}\right|^2} \cdot Re(Y_L) \qquad \text{[A3.13]}$$

$$P'_L = \frac{|b_{Sint}|^2}{|(1-\Gamma_L.\Gamma_S)|^2} \cdot (1 - |\Gamma_L|^2) \qquad \text{[A3.14]}$$

We see in expressions [A3.9] and [A3.13] a small defect of Mason's rule appears which is the following: the expressions are not necessarily natively in their most legible form. It becomes more legible to rewrite these two equations in the form [A3.15] for [A3.9] and [A3.16] for [A3.13]:

$$P_L = \frac{|E|^2|Y_{21}|^2 \cdot |Y_S|^2 \cdot |Y_L|^2}{|Y_S \cdot Y_L + Y_{11} \cdot Y_L + Y_{22} \cdot Y_S - Y_{21} \cdot Y_{12} + Y_{11} \cdot Y_{22}|^2} \cdot Re(1/Y_L)$$

$$= \frac{|E|^2|Y_{21}|^2 \cdot |Y_S|^2}{|Y_S \cdot Y_L + Y_{11} \cdot Y_L + Y_{22} \cdot Y_S - Y_{21} \cdot Y_{12} + Y_{11} \cdot Y_{22}|^2} \cdot Re(|Y_L|^2/Y_L)$$

$$P_L = \frac{|E|^2 \cdot |Y_{21}|^2 \cdot |Y_S|^2}{|Y_S \cdot Y_L + Y_{11} \cdot Y_L + Y_{22} \cdot Y_S - Y_{21} \cdot Y_{12} + Y_{11} \cdot Y_{22}|^2} \cdot Re(Y_L) \qquad \text{[A3.15]}$$

$$P'_L = \frac{|E|^2 \cdot |Y_S|^2}{|Y_S + Y_L|^2} \cdot Re(Y_L) \qquad \text{[A3.16]}$$

To calculate the power available at the source, it suffices to let $Y_L = Y_S^*$ in [A3.16] and $\Gamma_L = \Gamma_S^*$ in [A3.14]. Expressions [A3.17] and [A3.18] are respectively obtained:

$$P_{Savai} = \frac{|E|^2 \cdot |Y_S|^2}{|Y_S + Y_L|^2} \cdot Re(Y_L)\Big|_{Y_L = Y_S^*} = \frac{|E|^2 \cdot |Y_S|^2}{4.G_S} \quad \text{[A3.17]}$$

where G_S is the real part of Y_S.

$$P_{Savai} = \frac{|b_{Sint}|^2}{|(1 - \Gamma_L \cdot \Gamma_S)|^2} \cdot (1 - |\Gamma_L|^2)\Big|_{\Gamma_L = \Gamma_S^*} = \frac{|b_{Sint}|^2}{1 - |\Gamma_S|^2} \quad \text{[A3.18]}$$

We now have all the elements to express the transducer power gain of the two-port using equations [A3.9] and [A3.17] in Y parameters and [A3.10] and [A3.18] in S parameters. The expressions are given by equations [A3.19] and [A3.20]:

$$G_T(Y_L, Y_S) = \frac{4 \cdot G_S \cdot G_L \cdot |Y_{21}|^2}{|Y_S \cdot Y_L + Y_{11} \cdot Y_L + Y_{22} \cdot Y_S - Y_{21} \cdot Y_{12} + Y_{11} \cdot Y_{22}|^2} \quad \text{[A3.19]}$$

Where G_L is the real part of Y_L.

$$G_T(\Gamma_L, \Gamma_S) = \frac{|S_{21}|^2 \cdot (1 - |\Gamma_L|^2) \cdot (1 - |\Gamma_S|^2)}{|(1 - S_{11} \cdot \Gamma_S) \cdot (1 - S_{22} \cdot \Gamma_L) - S_{12} \cdot S_{21} \cdot \Gamma_S \cdot \Gamma_L|^2} \quad \text{[A3.20]}$$

To express the available power gain, the expressions of $Y_{Out} = Y_{22} - \frac{Y_{21} \cdot Y_{12}}{Y_S + Y_{11}}$ and of $\Gamma_{Out} = S_{11} + \frac{S_{21} \cdot S_{12} \cdot \Gamma_S}{1 - \Gamma_S \cdot S_{11}}$ must be shown in, respectively, equations [A3.19] and [A3.20] then let $Y_L = Y_{Out}^*$ and $\Gamma_L = \Gamma_{OUT}^*$. The expressions obtained are given by [A3.21] and [A3.22]:

$$G_{Avai}(Y_S) = \frac{4 \cdot G_S \cdot G_L \cdot |Y_{21}|^2}{|(Y_{Out} + Y_L) \cdot (Y_S + Y_{11})|^2}\Big|_{Y_L = Y_{Out}^*} = \frac{G_S \cdot |Y_{21}|^2}{G_{Out} \cdot |Y_S + Y_{11}|^2} \quad \text{[A3.21]}$$

where G_{Out} is the real part of Y_{Out}.

$$G_{Avai}(\Gamma_S) = \frac{|S_{21}|^2 \cdot (1 - |\Gamma_L|^2) \cdot (1 - |\Gamma_S|^2)}{|(1 - S_{11} \cdot \Gamma_S) \cdot (1 - \Gamma_{OUT} \cdot \Gamma_L)|^2}\Big|_{\Gamma_L = \Gamma_{Out}^*} = \frac{|S_{21}|^2 \cdot (1 - |\Gamma_S|^2)}{(1 - |\Gamma_{Out}|^2) \cdot |1 - S_{11} \cdot \Gamma_S|^2} \quad \text{[A3.22]}$$

To express the insertion power gains, it is enough to resume [A3.15] and [A3.16] in admittance and [A3.10] and [A3.14] in S parameters. The obtained expressions are given by [A3.23] and [A3.24]:

$$G_i(Y_L, Y_S) = \frac{|Y_{21}|^2 \cdot |Y_S + Y_L|^2}{|Y_S \cdot Y_L + Y_{11} \cdot Y_L + Y_{22} \cdot Y_S - Y_{21} \cdot Y_{12} + Y_{11} \cdot Y_{22}|^2} \quad \text{[A3.23]}$$

$$G_i(\Gamma_L, \Gamma_S) = \frac{|S_{21}|^2 \cdot |(1 - \Gamma_L \cdot \Gamma_S)|^2}{|(1 - S_{11} \cdot \Gamma_S) \cdot (1 - S_{22} \cdot \Gamma_L) - S_{12} \cdot S_{21} \cdot \Gamma_S \cdot \Gamma_L|^2} \quad \text{[A3.24]}$$

If we compare the insertion power gain and the available power gain in the case of a noise power measurement with a an adapted measuring chain and a noise diode adapted on $50\ \Omega(\Gamma_L = 0, \Gamma_D = 0)$, we find the error committed which is given by equation [A3.25]:

$$\frac{G_i(\Gamma_L, \Gamma_S)|_{\Gamma_L = 0, \Gamma_D = 0}}{G_{Avai}(\Gamma_S)|_{\Gamma_L = S_{22}^*, \Gamma_D = 0}} = \left. \frac{1}{(1 - |S_{22}|^2) \cdot (1 - |\Gamma_S|^2)} \right|_{\Gamma_S = 0} = \frac{1}{1 - |S_{22}|^2} \quad \text{[A3.25]}$$

It can thus be seen that if we take the measurement of noise factor on $50\ \Omega$, we do not make a mistake if the device under test is adapted to the output as mentioned in section 4.5.

A3.4. Conclusion and limits of Mason's rule

We explained how Mason's rule allows obtaining all the necessary quantities to know the operation of a two-port inserted between a generator and a load. Moreover, this is done without complicated calculations based on the writing of linear equations and elimination of variables.

We even showed that we could go from a representation of the two-port in S parameters to a representation in admittance parameters without having to redo calculations. This is the main advantage of using the flow graph and Mason's rule. The only disadvantage, already mentioned, is that sometimes it is necessary to reformulate the rough results of application of Mason's rule to obtain a more legible form.

Finally, we can say that in theory, we can apply Mason's rule to any flow graph. In practice, we are quickly limited by the human capacity to identify all the loops of a graph and the different paths as soon as this graph comprises more than a few tens of transfers t_{lm}. Nevertheless, four-port components, such as certain radiofrequency or microwave devices, can be analyzed in their environment, that is, with loads and generators, using this method.

For the record, we can show that the Δ_{Mason} is the determinant of the main matrix when we take the trouble to present the system of linear equations in the form of a system to be solved by the Cramer method as indicated by expression [A3.26]:

$$\begin{bmatrix} Main\ matrix\ (square) \\ describing\ the\ circuit's \\ constitution\ and \\ connections \end{bmatrix} \cdot \begin{bmatrix} Column \\ matrix \\ of\ created \\ quantities \end{bmatrix} = \begin{bmatrix} Column \\ matrix \\ of\ excitations \\ (causal \\ nodes) \end{bmatrix} \qquad \text{[A3.26]}$$

Appendix 4

Noise Power Waves

A4.1. Concept of noise power waves

The reason for the representation of noise in a two-port by two noise power waves comes from a concern of compatibility with the S parameters which represent relations between power waves. The most relevant representation (Heicken 1981) is given as a flow graph in Figure A4.1 and in matrix form by equation [A4.1].

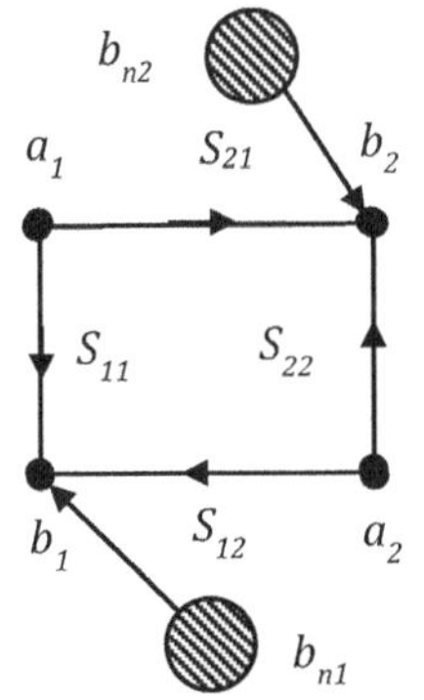

Figure A4.1. *Flow graph of a two-port by noise power waves*

$$\begin{vmatrix} b_1 \\ b_2 \end{vmatrix} = \begin{bmatrix} S_{11} & S_{12} \\ S_{21} & S_{22} \end{bmatrix} \cdot \begin{vmatrix} a_1 \\ a_2 \end{vmatrix} + \begin{vmatrix} b_{n1} \\ b_{n2} \end{vmatrix} \qquad [A4.1]$$

This representation has a particularity that if Γ_S the reflection coefficient of the source presented on port 1 of the two-port is zero, then the power wave b_{n1} does not contribute to the total noise collected at the output of the two-port. This does not call into question the relevance of the model because, as in the representation, presented in Chapter 1, generators b_{n1} and b_{n2} are equivalent generators, seen on the two-port accesses, to all that happens in terms of noise generation inside this same two-port. So we do not lose any information on the noise generated in the two-port when $\Gamma_S = 0$. It may be noted that, conceptually, it would be the same for i if $Z_S = 0$ and for e if $Y_S = 0$, even if these two situations, in practice, are highly improbable.

A4.2. Expression of the noise factor

To express the noise factor of a two-port using noise power waves, we will consider that this two-port is inserted between a source of reflection coefficient Γ_S and load of reflection coefficient Γ_L. The corresponding flow graph is given in Figure A4.2.

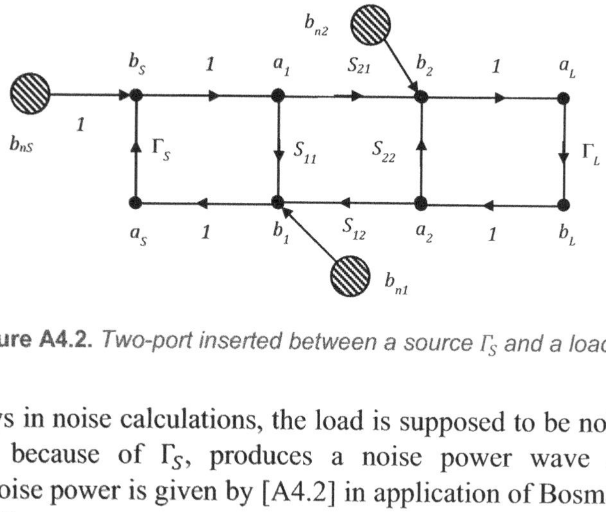

Figure A4.2. *Two-port inserted between a source Γ_S and a load Γ_L*

As always in noise calculations, the load is supposed to be noise free and the source, because of Γ_S, produces a noise power wave b_{nS} whose associated noise power is given by [A4.2] in application of Bosma's theorem (Bosma 1967):

$$\overline{|b_{nS}|^2} = kTdf \cdot (1 - |\Gamma_S|^2) \qquad \text{[A4.2]}$$

where T is the source temperature.

Using this representation, we can calculate $F(\Gamma_S)_{lin}$ the noise factor of the two-port by comparing the noise power collected in Γ_L taking into account b_{nS}, b_{n1} and b_{n2} and that collected in Γ_L by holding only count of b_{nS}. Since b_{nS} is not correlated to b_{n1} neither to b_{n2}, we can write the noise factor in the form given by [A4.3]:

$$F(\Gamma_S)_{lin} = 1 + \frac{(|a_L|^2 - |b_L|^2)|_{b_{nS}=0}}{(|a_L|^2 - |b_L|^2)|_{b_{n1}=b_{n2}=0}} \qquad \text{[A4.3]}$$

with $|b_L|^2 = |a_L|^2 \cdot |\Gamma_L|^2$.

Under these conditions, $F(\Gamma_S)_{lin}$ is rewritten in the form given by [A4.4] and does not depend on Γ_L; it is sufficient that the two noise powers are evaluated under the same load conditions:

$$F(\Gamma_S)_{lin} = 1 + \frac{|\overline{a_L}|^2|_{b_{nS}=0}}{|\overline{a_L}|^2|_{b_{n1}=b_{n2}=0}} \qquad \text{[A4.4]}$$

Then there is only the three transfers to calculate $\frac{a_L}{b_{nS}}$, $\frac{a_L}{b_{n1}}$ and $\frac{a_L}{b_{n2}}$. As the flow graph does not change for these three calculations, Δ_{Mason}, given by equation [A4.5], is the same for the three calculations and will be simplified in the expression of $F(\Gamma_S)_{lin}$. We nevertheless need its expression to calculate the Δ_{Path} that are different. It will be noted that the three loops are the same as on the flow graph of Figure A3.2:

$$(1 - S_{11} \cdot \Gamma_S) \cdot (1 - S_{22} \cdot \Gamma_L) - S_{12} \cdot S_{21} \cdot \Gamma_S \cdot \Gamma_L \qquad \text{[A4.5]}$$

For b_{nS}, the only direct path is $T_1 = S_{21}$ and it has at least one common transfer with the three loops, thus $\Delta_1 = 1$.

For b_{n1}, the only direct path is $T_2 = \Gamma_S \cdot S_{21}$ and it has at least one common transfer with the three loops, thus $\Delta_1 = 1$.

For b_{n2}, the only direct path is $T_3 = 1$ and it has at least one common transfer with loops B_2 and B_3 but not with B_1, thus $\Delta_3 = 1 - S_{11} \cdot \Gamma_S$.

We note here all the interest of the flow graph and Mason's rule which yield without difficulty the desired expressions.

We obtain for the noise factor the expression given by equation [A4.6]:

$$F(\Gamma_S)_{lin} = 1 + \frac{\left|\overline{\Gamma_S \cdot S_{21} \cdot b_{n1} + (1 - \Gamma_S \cdot S_{11}) \cdot b_{n2}}\right|^2}{|S_{21}|^2 \cdot |\overline{b_{nS}}|^2} \qquad \text{[A4.6]}$$

To develop the expression [A4.6] is not of interest at this stage of the computation, we will rather be interested in the b_{ni} to express their quadratic squared value $\overline{|b_{ni}|^2}$ and the correlation term $\overline{b_{ni}.b_{nj}^*}$.

Taking again the notations of Dobrowolski (1989), we use equation [A4.7]:

$$[C_S] = \begin{bmatrix} \overline{|b_{n1}|^2} & \overline{b_{n1}.b_{n2}^*} \\ \overline{b_{n1}^*.b_{n2}} & \overline{|b_{n2}|^2} \end{bmatrix} = kT.df \begin{bmatrix} C_{S11} & C_{S12} \\ C_{S12}^* & C_{S22} \end{bmatrix} =$$
$$kT.df \begin{bmatrix} c_{S11} & S_{21}^*.c_{S12} \\ S_{21}.c_{S12}^* & |S_{21}|^2.c_{S22} \end{bmatrix} \qquad \text{[A4.7]}$$

These slightly surprising notations, as to the form of the terms of the matrix c_S, will find their explanations in the following sections.

With these new notations, equation [A4.6] is written in the form [A4.8]:

$$F(\Gamma_S)_{lin} = 1 + \frac{|\Gamma_S|^2 \cdot c_{11} + |1-\Gamma_S \cdot S_{11}|^2 \cdot c_{22} + 2 \cdot Re[(1-\Gamma_S \cdot S_{11}) \cdot \Gamma_S^* \cdot c_{21}]}{1-|\Gamma_S|^2} \qquad \text{[A4.8]}$$

A4.3. Passage relations of b_{n1}, b_{n2} with e,i

To express these passage relations, we must work by using the representations of Figures A4.3 and A4.4. Figure A4.3 is the noise power waveform with terminations that make it easy to express b_{n1} and b_{n2}. Figure A4.4 is the same representation but in Y parameters and with the noise described by e, i. In these two representations, the terminal loads are Noise free because we are interested only in the noise generated in the two-port.

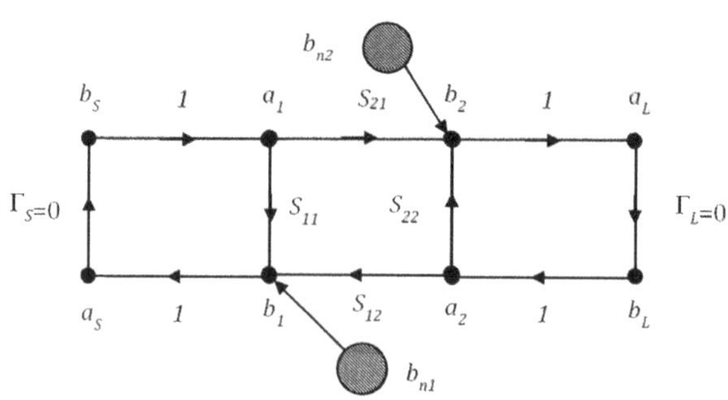

Figure A4.3. *Two-port inserted between a source $\Gamma_S = 0$ and a load $\Gamma_L = 0$*

Using the representation of Figure A4.3, thanks to $\Gamma_S = \Gamma_L = 0$, we can directly write the equalities [A4.9] and [A4.10] by using the passage relations between normalized current / voltage and power waves [A2.16]:

$$b_{n2} = a_L = b_2 = \frac{v_2 - i_2}{2} = v_2 = -i_2 \qquad \text{[A4.9]}$$

$$b_{n1} = a_S = b_1 = \frac{v_1 - i_1}{2} = v_1 = -i_1 \qquad \text{[A4.10]}$$

Using the representation of Figure A4.4, we can easily calculate V_1 and V_2 as a function of e and i.

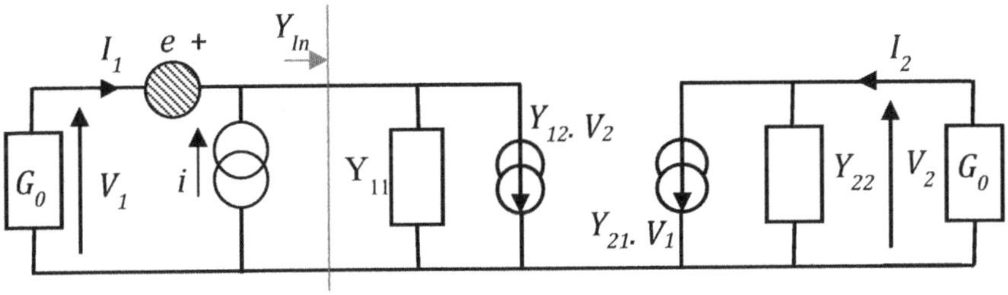

Figure A4.4. *Equivalent noise diagram of matrix two-port admittance [Y] loaded at the input and output by* G_0

In this circuit, the admittance seen on port 1 of the two-port is $Y_{In} = Y_{11} - \frac{Y_{21}.Y_{12}}{G_0 + Y_{22}}$ as defined in section A3.3.2 of Annex 3 for an output load G_0.

According to [A2.17], we can write $S_{11} = \frac{Z_{IN1}|_{R_0} - R_0}{Z_{IN1}|_{R_0} + R_0} = \frac{G_0 - Y_{In}}{G_0 - Y_{In}}$ with $G_0 = 1/R_0$ thus, $Y_{In} = G_0.\frac{1 - S_{11}}{1 + S_{11}}$ or $\frac{Y_{In}}{G_0} = y_{In} = \frac{1 - S_{11}}{1 + S_{11}}$.

To calculate $-I_1$(or V_1) without making any unnecessary calculations, we replace the input set e, i, Y_{In}, seen from port 1, by its equivalent voltage generator $e_{éq1} = -e + \frac{i}{Y_{In}}$ and its equivalent admittance Y_{In}. The corresponding arrangement is given in Figure A4.5.

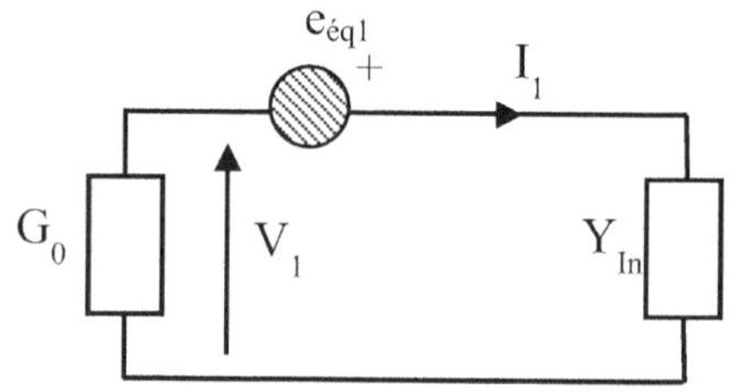

Figure A4.5. *Simplified two-port noise equivalent diagram for the calculation of $-I_1$ or V_1*

The expression for current I_1 is thus $I_1 = -\frac{G_0 \cdot Y_{In} \cdot e_{éq1}}{G_0 + Y_{In}} = \frac{G_0 \cdot Y_{In} \cdot e}{G_0 + Y_{In}} - \frac{G_0 . i}{G_0 + Y_{In}} = \frac{Y_{In} . e}{1 + Y_{In}/G_0} - \frac{i}{1 + Y_{In}/G_0}$.

The expression for the normalized current $i_1 = I_1/\sqrt{G_0}$ becomes [A4.11]:

$$i_1 = \frac{y_{In}}{1+y_{In}} . e_{Nor} - \frac{1}{1+y_{In}} . i_{Nor} \qquad [A4.11]$$

If we replace y_{In} by its expression as a function of S_{11}, we obtain equation [A4.12]:

$$b_{n1} = \frac{S_{11}-1}{2} . e_{Nor} + \frac{S_{11}+1}{2} . i_{Nor} \qquad [A4.12]$$

with $i_{Nor} = \frac{i}{\sqrt{G_0}}$ and $e_{Nor} = e.\sqrt{G_0}$.

To calculate V_2(or $-I_2$) without making any unnecessary calculations, we replace the input set e, i, G_0, seen from Y_{In}, by its equivalent voltage generator $e_{éq2} = e + \frac{i}{G_0}$ and its equivalent admittance G_0. The corresponding arrangement is given in Figure A4.6.

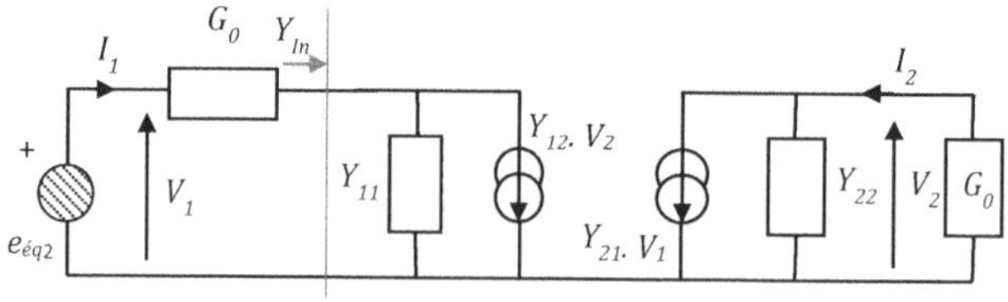

Figure A4.6. *Simplified two-port noise equivalent diagram for the calculation of V_2 or $-I_2$*

The ratio $\frac{V_2}{e_{éq2}}$ is equal to the composite gain of the two-port for source and load admittances equal to G_0. We have seen in Annex 2 that this quantity is equal to $\frac{S_{21}}{2}$. Thus the expression for V_2 is written in the form given by equation [A4.13]:

$$V_2 = \frac{S_{21}}{2} \cdot \left[e + \frac{i}{G_0}\right] \quad [A4.13]$$

After normalization with respect to R_0, we obtain the expression [A4.14]:

$$b_{n2} = \frac{S_{21}}{2} \cdot [e_{Nor} + i_{Nor}] \quad [A4.14]$$

with $e_{Nor} = \frac{e}{\sqrt{R_0}}$ and $i_{Nor} = i \cdot \sqrt{R_0}$.

A4.4. Noise power waves correlation matrix

We will now express the c_{Sij} of the matrix $[C_S]$ as written in [A4.7] by using [A4.12] and [A4.14].

For c_{22}, we obtain by identification of [A4.14] expression [A4.15]:

$$c_{S22} = r_N \cdot |1 + y_{Cor}|^2 + g_n \quad [A4.15]$$

We therefore need to express c_{S22} as a function of the usual noise parameters which are the minimum noise factor F_{Min}, the equivalent noise resistance at the input of the two-port in normalized value r_N and the reflection coefficient of the optimum source for the noise Γ_{Sopt} or the optimal source admittance for the Y_{Sopt} noise. We will also use equations [1.14] and [1.15] which are written in normalizd values to $R_0 = 50\ \Omega$ for more convenience:

$$y_{Sopt} = g_{Sopt} + jb_{Sopt} = \sqrt{g_{Cor}^2 + \frac{g_n}{r_N}} - jb_{Cor} \quad [1.14]$$

$$F_{Min} = 1 + 2 \cdot r_N \cdot (g_{Sopt} + g_{Cor}) \quad [1.15]$$

$$c_{S22} = r_N \cdot |1 + y_{Cor}|^2 + g_n$$

$$= r_N \cdot [1 + g_{Cor}{}^2 + 2 \cdot g_{Cor}] + g_n + r_N \cdot b_{Cor}{}^2$$

$$= r_N \cdot \left[1 + g_{Cor}{}^2 + \frac{g_n}{r_N}\right] + 2 \cdot r_N \cdot g_{Cor} + r_N.b_{Cor}{}^2$$

$$= r_N \cdot \left[1 + g_{Sopt}{}^2\right] + 2 \cdot r_N \cdot g_{Cor} + r_N.b_{Cor}{}^2$$

$$= r_N \cdot \left[1 + g_{Sopt}{}^2 - 2 \cdot g_{Sopt}\right] + 2 \cdot r_N \cdot \left(g_{Cor} + g_{Sopt}\right) + r_N.b_{Cor}{}^2$$

$$= r_N \cdot \left[1 - g_{Sopt}\right]^2 + r_N.b_{Sopt}{}^2 + 2 \cdot r_N \cdot \left(g_{Cor} + g_{Sopt}\right)$$

$$c_{S22} = r_N \cdot \left|1 - y_{Sopt}\right|^2 + F_{Min} - 1 = r_N \cdot \left|1 - y_{Sopt}\right|^2 + F_{eMin} \qquad \text{[A4.16]}$$

If y_{Sopt} is known, the most practical expression of c_{S22} is that given by [A4.16] where the quantity F_{eMin} is the minimum excess noise factor of the two-port equal to the minimum noise factor of the two-port minus 1.

The result of [A4.16] can be obtained more simply using [A4.8] by making $\Gamma_S = 0$; giving $F(\Gamma_S = 0)_{lin} = 1 + c_{S22}$. At the same time, using [1.16], we get, for $y_S = 1$ which corresponds to $\Gamma_S = 0$, $F(y_S = 1) = F_{Min} + r_N \cdot \left|1 - y_{Sopt}\right|^2$. This confirms the expression for c_{S22} given by [A4.16].

If we wish to work with Γ_{Sopt}, we use $y_{Sopt} = \frac{1-\Gamma_{Sopt}}{1+\Gamma_{Sopt}}$ in [A4.16] to obtain [A4.17]. This form is proposed by (Wedge *et al.* 1992) and it uses T_{eMin} the minimum excess temperature of the two-port given by $T_{eMin} = F_{eMin} \cdot T_0$:

$$c_{S22} = 4 \cdot r_N \cdot \frac{\left|\Gamma_{Sopt}\right|^2}{\left|1+\Gamma_{Sopt}\right|^2} + F_{eMin} \qquad \text{[A4.17]}$$

This expression is also confirmed using [1.17a] for $\Gamma_S = 0$, corresponding to $y_S = 1$, we obtain $F(\Gamma_S = 0) = F_{Min} + 4 \cdot r_N \cdot \frac{\left|\Gamma_{Sopt}\right|^2}{\left|1+\Gamma_{Sopt}\right|^2}$.

We can also write this equation in the form given by equation [A4.18]:

$$c_{S22} = \frac{F_{eMin} + \left|\Gamma_{Sopt}\right|^2 \cdot \left[4 \cdot r_N \cdot g_{Sopt} - F_{eMin}\right]}{1 - \left|\Gamma_{Sopt}\right|^2} \qquad \text{[A4.18]}$$

The form [A4.18], proposed by (Dobrowolski 1989), is not more practical to use than the form [A4.17], because we replace the calculation of $\left|1+\Gamma_{Sopt}\right|^2$ by that of g_{Sopt}.

For the calculation of c_{S11}, we will start from the expression [A4.12] giving the expression of b_{n1}. This expression can be rewritten in the form given by [A4.19]:

$$b_{n1} = \frac{S_{11}}{2}\cdot(e_{Nor} + i_{Nor}) - \frac{1}{2}\cdot(e_{Nor} - i_{Nor}) \qquad \text{[A4.19]}$$

For $|\overline{b_{n1}}|^2$, we obtain the following expression:

$$|\overline{b_{n1}}|^2 = \frac{1}{4}\cdot\left[|S_{11}|^2\cdot|\overline{e_{Nor} + \iota_{Nor}}|^2 + |\overline{e_{Nor} - \iota_{Nor}}|^2 \right.$$
$$\left. - 2\cdot Re(S_{11}\cdot\overline{(e_{Nor} + \iota_{Nor})\cdot(e^*_{Nor} - \iota^*_{Nor})})\right]$$

The term $|\overline{e_{Nor} + \iota_{Nor}}|^2$ is known for having calculated it to expressc_{S22}; it is equal to $4kT\cdot df\cdot(4\cdot r_N\cdot\frac{|\Gamma_{Sopt}|^2}{|1+\Gamma_{Sopt}|^2} + F_{eMin})$. We need to calculate the other two terms:

$$|\overline{e_{Nor} - \iota_{Nor}}|^2 = \left|\overline{e_{Nor}(1-y_{Cor}) - \iota_{nNor}}\right|^2 = 4kT\cdot df\left[r_N\cdot(1+g_{Cor}{}^2 + b_{Cor}{}^2 - 2\cdot g_{Cor}) + g_n\right] = 4kT\cdot df\left[r_N\cdot\left(1+g_{Sopt}{}^2 + b_{Sopt}{}^2 - 2\cdot g_{Cor}\right)\right] = 4kT\cdot df\cdot r_N\cdot\left[(1+g_{Sopt})^2 + b_{Sopt}{}^2 - 2(g_{Sopt}+g_{Cor})\right] = 4kT\cdot df\cdot\left[-F_{eMin} + r_N\cdot\left|1+y_{Sopt}\right|^2\right] = 4kT\cdot df\cdot\left[-F_{eMin} + \frac{4r_N}{\left|1+\Gamma_{Sopt}\right|^2}\right]$$

$$\overline{(e_{Nor} + \iota_{Nor})\cdot(e^*_{Nor} - \iota^*_{Nor})} = 4kT\cdot df\left[r_N\cdot(1+y_{Cor})\cdot(1-y^*_{Cor}) - g_n\right] = 4kT\cdot df\cdot r_N\cdot\left(1-y_{Sopt}\right)\cdot\left(1+y^*_{Sopt}\right)$$

$$2\cdot Re\left(S_{11}\cdot\overline{(e_{Nor} + \iota_{Nor})\cdot(e^*_{Nor} - \iota^*_{Nor})}\right)$$

$$= 8kT\cdot df\cdot r_N\cdot Re(S_{11}\cdot\frac{(1-y_{Sopt})}{(1+y_{Sopt})}\cdot\left(1+y^*_{Sopt}\right)\cdot\left(1+y_{Sopt}\right))$$
$$= 8kT\cdot df\cdot r_N\cdot Re(S_{11}\cdot\frac{(1-y_{Sopt})}{(1+y_{Sopt})}\cdot\left|1+y_{Sopt}\right|^2) = 8kT\cdot df\cdot r_N\cdot\left|1+y_{Sopt}\right|^2\cdot Re\left(S_{11}\cdot\Gamma_{Sopt}\right) = 8kT\cdot df\cdot r_N\cdot\frac{4}{\left|1+\Gamma_{Sopt}\right|^2}\cdot Re\left(S_{11}\cdot\Gamma_{Sopt}\right)$$

We can therefore express $|\overline{b_{n1}}|^2$ and obtain the following expression:

$$|\overline{b_{n1}}|^2 = \frac{4kT.df}{4}.\left[|S_{11}|^2.\left(4.r_N.\frac{|\Gamma_{Sopt}|^2}{|1+\Gamma_{Sopt}|^2}+F_{eMin}\right) + \left[-F_{eMin}+\frac{4r_N}{|1+\Gamma_{Sopt}|^2}\right] - \frac{8.r_N}{|1+\Gamma_{Sopt}|^2}.Re(S_{11}.\Gamma_{Sopt})\right]$$

By identification, we obtain the expression of c_{11} given by [A4.20]:

$$c_{S11} = F_{eMin}(|S_{11}|^2 - 1) + \frac{4\cdot r_N\cdot|1-S_{11}\cdot\Gamma_{Sopt}|^2}{|1+\Gamma_{Sopt}|^2} \qquad \text{[A4.20]}$$

For the term $\overline{b_{n1}.b_{n2}^*}$ we obtain, with the same method, the following expression:

$$\overline{b_{n1}.b_{n2}^*} = \overline{\left[\frac{S_{11}}{2}\cdot(e_{Nor}+\iota_{Nor}) - \frac{1}{2}\cdot(e_{Nor}-\iota_{Nor})\right]\cdot\left[\frac{S_{21}^*}{2}\cdot[e_{Nor}^*+\iota_{Nor}^*]\right]}$$

$$\overline{b_{n1}.b_{n2}^*} = \frac{S_{11}\cdot S_{21}^*}{4}\cdot|\overline{e_{Nor}+\iota_{Nor}}|^2 - \frac{S_{21}^*}{4}\cdot\overline{(e_{Nor}-\iota_{Nor})\cdot(e_{Nor}^*+\iota_{Nor}^*)}$$

The first term is:

$$4kT\cdot df\cdot\frac{S_{11}\cdot S_{21}^*}{4}\cdot\left(4\cdot r_N\cdot\frac{|\Gamma_{Sopt}|^2}{|1+\Gamma_{Sopt}|^2}+F_{eMin}\right)$$

The second term is:

$$4kT\cdot df\cdot r_N\cdot\frac{S_{21}^*}{4}\cdot(1+y_{Sopt})\cdot(1-y_{Sopt}^*)$$

$$= 4kT\cdot df\cdot r_N\cdot\frac{S_{21}^*}{4}\cdot\frac{4}{|1+\Gamma_{Sopt}|^2}\cdot\Gamma_{Sopt}^*$$

$$\overline{b_{n1}.b_{n2}^*} = kT\cdot df\cdot S_{21}^*\cdot\left[S_{11}\cdot\left(4\cdot r_N\cdot\frac{|\Gamma_{Sopt}|^2}{|1+\Gamma_{Sopt}|^2}+F_{eMin}\right) - \frac{4\cdot r_N\cdot\Gamma_{Sopt}^*}{|1+\Gamma_{Sopt}|^2}\right]$$

By identification, we obtain c_{12} given by the expression [A4.21]:

$$c_{S12} = S_{11} \cdot \left(4 \cdot r_N \cdot \frac{|\Gamma_{Sopt}|^2}{|1+\Gamma_{Sopt}|^2} + F_{eMin}\right) - \frac{4 \cdot r_N \cdot \Gamma_{Sopt}^*}{|1+\Gamma_{Sopt}|^2} \qquad \text{[A4.21]}$$

The term c_{21} is obtained by taking the conjugate of c_{12}, its expression is given by equation [A4.22]:

$$c_{S21} = S_{11}^* \cdot \left(4 \cdot r_N \cdot \frac{|\Gamma_{Sopt}|^2}{|1+\Gamma_{Sopt}|^2} + F_{eMin}\right) - \frac{4 \cdot r_N \cdot \Gamma_{Sopt}}{|1+\Gamma_{Sopt}|^2} \qquad \text{[A4.22]}$$

Using [A4.18], [A4.20] and [A4.22] in [A4.8], we find, after some lines of calculations, the expression of the noise factor for a source reflection coefficient Γ_S identical to that given by equation [1.17a].

A4.5. Inverse relations

It is now necessary to express the noise parameters F_{Min}, r_N and Γ_{Sopt} as a function of the power wave correlation matrix which is accessible by measurements (Wedge *et al.* 1992) which will not be discussed here.

To establish these relations, one starts from the expression [A4.8] which gives the noise factor of the two-port F for a reflection coefficient of source Γ_S, of parameters S_{11}, S_{21} and terms c_{S11}, c_{S22} and c_{S21}. We can rewrite this relation in the form given by equation [A4.23]:

$$|\Gamma_S|^2 - \frac{2 \cdot Re[\Gamma_S(S_{11}c_{S22} - c_{S12})]}{c_{S11} + c_{S22} \cdot |S_{11}|^2 + F - 1 - 2Re(S_{11}^* \cdot c_{S12})}$$

$$+ \frac{c_{S22}+1-F}{c_{S11}+c_{S22} \cdot |S_{11}|^2+F-1-2Re(S_{11}^* \cdot c_{S12})} = 0 \qquad \text{[A4.23]}$$

This equation is that of a circle of center Ω and radius R in the complex plane as given by [A4.24]:

$$|\Gamma - \Omega|^2 - R^2 = |\Gamma|^2 - 2Re(\Gamma\Omega^*) + |\Omega|^2 - R^2 \qquad \text{[A4.24]}$$

Equation [A4.23] reflects the fact that a set of Γ_S produces the same noise factor F.

Since c_{S22} is real because it is the mean squared value of b_{n2} with the coefficient $kT \cdot df \cdot |S_{21}|^2$ the equation [A4.23] corresponds to a circle of center Ω whose expression is given by [A4.25]:

$$\Omega = \frac{S_{11}^* \cdot c_{S22} - c_{S12}^*}{c_{S11} + c_{S22} \cdot |S_{11}|^2 + F - 1 - 2Re(S_{11}^* \cdot c_{S12})} \qquad \text{[A4.25]}$$

When the noise factor F is equal to F_{Min}, this set is reduced to a single point which corresponds to $\Omega = \Gamma_{Sopt}$ and the radius R is zero. This results in the equality given by [A4.26]:

$$\frac{|S_{11}^* \cdot c_{S22} - c_{S12}^*|^2}{[c_{S11} + c_{S22} \cdot |S_{11}|^2 + F_{eMin} - 2Re(S_{11}^* \cdot c_{S12})]^2} = \frac{c_{22} - F_{eMin}}{c_{S11} + c_{S22} \cdot |S_{11}|^2 + F_{eMin} - 2Re(S_{11}^* \cdot c_{S12})} \qquad \text{[A4.26]}$$

This means that F_{eMin} is the solution of a second degree equation given by [A4.27]:

$$F_{eMin}^{\ 2} - F_{eMin}\left(c_{S22}(1 - |S_{11}|^2) - c_{S11} + 2Re(S_{11}^* \cdot c_{S12})\right) + |c_{S12}|^2 - c_{S11} \cdot c_{S22} = 0 \qquad \text{[A4.27]}$$

To solve this equation, it should be remembered that the minimum excess noise factor of a two-port is unique and strictly positive.

Equation [A4.27] can be rewritten as $x^2 + ax + b = 0$ where $-$a is the sum of the roots and b is their product. Using Table A4.1, we can find the case where we have a unique positive root.

Sum	Product	Roots	Solution
$-a > 0$	$b > 0$	2 roots> 0	Not possible since F_{eMin} unique
$-a > 0$	$b < 0$	$\frac{-a+\sqrt{\Delta}}{2} > 0$ et $\frac{-a-\sqrt{\Delta}}{2} < 0$	$\frac{-a+\sqrt{\Delta}}{2}$
$-a < 0$	$b > 0$	2 roots< 0	No solutions since $F_{eMin} > 0$
$-a < 0$	$b < 0$	$\frac{-a+\sqrt{\Delta}}{2} > 0$ et $\frac{-a-\sqrt{\Delta}}{2} < 0$	$\frac{-a+\sqrt{\Delta}}{2}$

Table A4.1. *Positive root(s) of a second degree equation*

The only appropriate case is b<0 and we take the root $\frac{-a+\sqrt{\Delta}}{2}$. We can note that b<0 means that the term $c_{S11} \cdot c_{S22} - |c_{S12}|^2$ is positive.

Physically, it is very understandable if we reason on the correlation between b_{n1} and b_{n2} as shall be demonstrated.

Let $b_{n1} = b_{n1déc} + c \cdot b_{n2}$ where $b_{n1déc}$ is part of b_{n1}that is not correlated to b_{n2} and c a correlation coefficient between $b_{n1}$and b_{n2}. With these notations, we can express $\overline{b_{n1}.b_{n2}^*}$ in the form $\overline{b_{n1}.b_{n2}^*} = kT \cdot df \cdot S_{21}^* \cdot c_{S12} = kT \cdot df \cdot c \cdot |S_{21}|^2 \cdot c_{S22}$This means that $c = \frac{c_{S12}}{S_{21} \cdot c_{S22}}$. In other words, we can write that $|\overline{b_{n1}}|^2 = kT \cdot df \cdot c_{S11} = kT \cdot df \cdot (c_{S11déc} + |c|^2 \cdot |S_{21}|^2 \cdot c_{S22}) = kT \cdot df \cdot \left(c_{S11déc} + \frac{|c_{S12}|^2}{c_{S22}}\right)$ where $|b_{n1déc}|^2 = kT \cdot df \cdot c_{S11déc}$. We thus obtain $c_{S11} \cdot c_{S22} = c_{S11déc} \cdot c_{S22} + |c_{S12}|^2$. We see that $c_{S11} \cdot c_{S22} - |c_{S12}|^2$ is positive; this term is zero only if b_{n1} and b_{n2} are totally correlated and $c_{S11déc}=0$.

Thus, we can express F_{eMin} as given by equation [A4.28]:

$$F_{eMin} = \frac{c_{S22}(1-|S_{11}|^2) - c_{S11} + 2Re(S_{11}^* \cdot c_{S12})}{2} \cdot \left(1 - \sqrt{1 - \frac{4 \cdot (|c_{S12}|^2 - c_{S11} \cdot c_{S22})^2}{[c_{S22}(1-|S_{11}|^2) - c_{S11} + 2Re(S_{11}^* \cdot c_{S12})]^2}}\right) \qquad [A4.28]$$

To find Γ_{Sopt}, we report the calculated value F_{eMin} in equation [A4.25] to find equation [A4.29]:

$$\Gamma_{Sopt} = \frac{S_{11}^* \cdot c_{S22} - c_{S12}^*}{c_{S11} + c_{S22} \cdot |S_{11}|^2 + F_{eMin} - 2Re(S_{11}^* \cdot c_{S12})} \qquad [A4.29]$$

Finally, r_N is obtained using equation [A4.17] rewritten in the form equation [A4.30]:

$$r_N = (c_{S22} - F_{eMin}) \cdot \frac{|1+\Gamma_{Sopt}|^2}{4 \cdot |\Gamma_{Sopt}|^2} \qquad [A4.30]$$

A4.6. Conclusion

The only drawback of this formulation resides in the establishment of the formulas of direct and inverse passage between the "classical" noise parameters which are F_{Min}, r_N and Γ_{Sopt} and the correlation matrix of the noise power waves. However, once this work is done, the use of noise power waves as a representation of noise in a two-port, although still less widespread than the others, is particularly interesting in the following situations:

– calculating the noise characteristics of a two-port resulting from any combination of devices having any number of ports but known by their S matrix and their noise power wave correlation matrix (Dobrowolski 1989). This is made possible thanks to a matrix formulation of the passage relations between noise generators, in voltage or current, on the different accesses of the devices and the resulting noise power waves. This technique also uses the connection matrix describing how the different devices are connected to each other. This is also the method used in the QUCS simulation software;

– noise factor measurement with a vector network analyzer. Some of these analyzers propose a method based on noise power waves to measure two-port noise characteristics without the use of a tuner or noise source. These analyzers measure the correlation matrix of the noise power waves and deduce the quantities F_{Min}, r_N and Γ_{Sopt}. As for the "classical" measurement, it is important to make a calibration to eliminate the noise contribution of the two receivers inside the network analyzer that measure b_{n1} and b_{n2}. The use of external low-noise amplifiers is also necessary to mask the noise factor of these two receivers which is generally of the order of about thirty dB. This method can be used with heterodyne vector network analyzers, but also six-port type.

Bibliography

Anderson, R.W. (1967). S-Parameter techniques for faster, more accurate network design. *HP Journal*, 18(6), 13–22.

Asgaram, S., Deen, M.J., Chen, C. (2004). Analytical modeling of MOSFETs channel noise and noise parameters. *IEEE Trans. on Electrons Devices*, 51(12), 2109–2114.

Bosma, H. (1967). On theory of linear noisy systems. *Philips Research Supplements*, 10.

Chen, C., Deen, M.J., Cheng, Y. (2001). Extraction of the induced gate noise, chanel noise and their correlation in submicron MOSFETs from RF noise measurements. *IEEE TED*, 48(12), 2884–2892.

Dobrowolski, J.A. (1989). A CAD-oriented method for noise figure computation of two-ports with any internal topology. *IEEE Trans. Microwave Theory and Techniques*, MTT-37(1), 194–202.

Friis, H.T. (1944). Noise figures of radio receivers. *Proc. IRE*, 32, 419–422.

Gao, J. (2011). Direct parameter-extraction method for MOSFET noise model from microwave noise figure measurement. *Solid-State Electronics*, 63, 42–48.

Heiken, R.P. (1981). Analysis of linear two-ports using scattering waves. *IEEE Trans. Microwave Theory and Techniques*, MTT-29(10), 997–1004.

Keysight Technologies (2017a). Fundamentals of RF and microwave noise figure measurements [Online]. Available at: http://literature.cdn.keysight.com/litweb/pdf/5952-8255E.pdf, accessed November 2018.

Keysight Technologies (2017b). Noise figure uncertainty calculator spreadsheet [Online]. Available at: https://www.keysight.com/main/editorial.jspx%3Fcc%3DUS%26lc%3Deng%26ckey%3D96887%26nid%3D-11143.0.00%26id%3D96887%29?&cc=GB&lc=eng, accessed November 2019.

Kurokawa, K. (1965). Power waves and scattering matrix. *IEEE Trans. Microwave Theory and Techniques*, MTT-13(2), 194–202.

Mason, S.J. (1956). Feedback theory – further properties of signal flow graphs. *Proc. IRE*, 44(7), 920–926.

Mini-Circuits (2019). Datasheet du ZHL-1010+ [Online]. Available at: https://ww2.minicircuits.com/pdfs/ZHL-1010.pdf, accessed April 2019.

Rohde & Schwarz (2018). The Y factor technique for noise figure measurements [Online]. Application note. Available at: https://www.rohde-schwarz-ad.com/wp-content/uploads/2018/07/1MA178_3e_NoiseFigure.pdf, accessed November 2019.

Rota, F. (2007). Noise source diodes. *VHF Communications*, 4, 12–18.

Rothe, H., Dalke, W. (1956). Theory of noisy four poles. *Proc. IRE*, 44, 811–818.

Van der Ziel, A. (1963). Gate noise in field effect transistors at moderately high frequencies. *Proc. IRE*, 51, 461–467.

Vendelin, G.D., Pavio, A.M., Rohde, U.L. (2005). *Microwave Circuit Design Using Linear and Nonlinear Techniques*, 2nd edition. Wiley, New York.

Wedge, S.W., Rutledge, D.B. (1992). Wave techniques for noise modeling and measurement. *IEEE Trans. Microwave Theory and Techniques*, MTT-40(11), 2004–2012.

Index

P

R

S, T, V

Other titles from

iSTE

in

Electronics Engineering

2018

MURET Pierre
Fundamentals of Electronics 2: Continuous-time Signals and Systems
Fundamentals of Electronics 3: Discrete-time Signals and Systems, and Quantized Level Systems

2017

MURET Pierre
Fundamentals of Electronics 1: Electronic Components and Elementary Functions

BUCCI Davide
Analog Electronics for Measuring Systems

2016

BAUDRAND Henri, TITAOUINE Mohammed, RAVEU Nathalie
The Wave Concept in Electromagnetism and Circuits: Theory and Applications

FANET Hervé
Ultra Low Power Electronics and Adiabatic Solutions

NDJOUNTCHE Tertulien
Digital Electronics 1: Combinational Logic Circuits
Digital Electronics 2: Sequential and Arithmetic Logic Circuits
Digital Electronics 3: Finite-state Machines

2015

DURAFFOURG Laurent, ARCAMONE Julien
Nanoelectromechanical Systems

2014

APPRIOU Alain
Uncertainty Theories and Multisensor Data Fusion

CONSONNI Vincent, FEUILLET Guy
Wide Band Gap Semiconductor Nanowires 1: Low-Dimensionality Effects and Growth
Wide Band Gap Semiconductor Nanowires 2: Heterostructures and Optoelectronic Devices

GAUTIER Jean-Luc
Design of Microwave Active Devices

LACAZE Pierre Camille, LACROIX Jean-Christophe
Non-volatile Memories

TEMPLIER François
OLED Microdisplays: Technology and Applications

THOMAS Jean-Hugh, YAAKOUBI Nourdin
New Sensors and Processing Chain

2013

COSTA François, GAUTIER Cyrille, LABOURE Eric, REVOL Bertrand
Electromagnetic Compatibility in Power Electronics

KORDON Fabrice, HUGUES Jérôme, CANALS Agusti, DOHET Alain
Embedded Systems: Analysis and Modeling with SysML, UML and AADL

LE TIEC Yannick
Chemistry in Microelectronics

2012

BECHERRAWY Tamer
Electromagnetism: Maxwell Equations, Wave Propagation and Emission

LALAUZE René
Chemical Sensors and Biosensors

LE MENN Marc
Instrumentation and Metrology in Oceanography

SAGUET Pierre
Numerical Analysis in Electromagnetics: The TLM Method

2011

ALGANI Catherine, RUMELHARD Christian, BILLABERT Anne-Laure
Microwaves Photonic Links: Components and Circuits

BAUDRANT Annie
Silicon Technologies: Ion Implantation and Thermal Treatment

DEFAY Emmanuel
Integration of Ferroelectric and Piezoelectric Thin Films: Concepts and Applications for Microsystems

DEFAY Emmanuel
Ferroelectric Dielectrics Integrated on Silicon

BESNIER Philippe, DÉMOULIN Bernard
Electromagnetic Reverberation Chambers

LANDIS Stefan
Nano-lithography

2010

LANDIS Stefan
Lithography

PIETTE Bernard
VHF / UHF Filters and Multicouplers

2009

DE SALVO Barbara
Silicon Non-volatile Memories / Paths of Innovation

DECOSTER Didier, HARARI Joseph
Optoelectronic Sensors

FABRY Pierre, FOULETIER Jacques
Chemical and Biological Microsensors / Applications in Fluid Media

GAUTIER Jacques
Physics and Operation of Silicon Devices in Integrated Circuits

MOLITON André
Solid-State Physics for Electronics

PERRET Robert
Power Electronics Semiconductor Devices

SAGUET Pierre
Passive RF Integrated Circuits

2008

CHARRUAU Stéphane
Electromagnetism and Interconnections

2007

RIPKA Pavel, TIPEK Alois
Modern Sensors Handbook

Printed and bound by CPI Group (UK) Ltd, Croydon, CR0 4YY
15/09/2026
14962352-0006